Synthese Library

Studies in Epistemology, Logic, Methodology, and Philosophy of Science

Volume 405

The aim of *Synthese Library* is to provide a forum for the best current work in the methodology and philosophy of science and in epistemology. A wide variety of different approaches have traditionally been represented in the Library, and every effort is made to maintain this variety, not for its own sake, but because we believe that there are many fruitful and illuminating approaches to the philosophy of science and related disciplines.

Special attention is paid to methodological studies which illustrate the interplay of empirical and philosophical viewpoints and to contributions to the formal (logical, set-theoretical, mathematical, information-theoretical, decision-theoretical, etc.) methodology of empirical sciences. Likewise, the applications of logical methods to epistemology as well as philosophically and methodologically relevant studies in logic are strongly encouraged. The emphasis on logic will be tempered by interest in the psychological, historical, and sociological aspects of science.

Besides monographs *Synthese Library* publishes thematically unified anthologies and edited volumes with a well-defined topical focus inside the aim and scope of the book series. The contributions in the volumes are expected to be focused and structurally organized in accordance with the central theme(s), and should be tied together by an extensive editorial introduction or set of introductions if the volume is divided into parts. An extensive bibliography and index are mandatory.

More information about this series at http://www.springer.com/series/6607

Preface

This volume originates from the second *Conceptual Spaces at Work* conference, initiated by Mauri Kaipainen and Antti Hautamäki, held August 24–27, 2016 at Södertörn University, Sweden. The book collects manuscripts received in response to an open call for papers on themes discussed at this event. We thank the Swedish Research Council and the Volkswagen Foundation for supporting the conference and a pre-conference workshop, respectively. Last but not least, we gratefully acknowledge the prompt and constructive service of a number of anonymous reviewers, without whom this book would not have been possible.

Helsinki, Finland	Mauri Kaipainen
Warsaw, Poland	Frank Zenker
Jyväskylä, Finland	Antti Hautamäki
Lund, Sweden	Peter Gärdenfors
May 2019	

Contents

Chapter 1
Editors' Introduction

Mauri Kaipainen, Frank Zenker, Antti Hautamäki, and Peter Gärdenfors

Unifying the manuscripts collected in this volume is Gärdenfors's (2000) theory of conceptual spaces. It has meanwhile established itself both within contemporary cognitive science and beyond, as a descriptive approach mediating between the symbolic and the sub-symbolic levels of representation.

One on hand, we here present more sophisticated applications of it, but also theoretical extensions, adaptations, and augmentations, on the other. Throughout the book, we can discern not only aspects of such developments (if at times only implicitly), there also is new evidence that the theory's empirical content extends beyond what its original assumptions had suggested. Thus, the volume generally speaks to the theory's maturity. At the same time, as the limits of the theory's validity and application-range become clearer, its identity is prone to change—a development we explicitly welcome.

The book's content is admittedly abstract and sometimes technical. It should nevertheless be of relevance to scholars in a broad range of areas, from philosophy via linguistics to AI. We hope that particularly readers outside of cognitive science and its neighboring areas find much that is useful to them. Though the papers are

M. Kaipainen
Perspicamus LTD, Helsinki, Finland
e-mail: mauri.kaipainen@perspicamus.com

F. Zenker (✉)
International Center for Formal Ontology, Warsaw University of Technology, Warsaw, Poland
e-mail: frank.zenker@fil.lu.se

A. Hautamäki
Department of Social Sciences and Philosophy, University of Jyväskylä, Jyväskylä, Finland

P. Gärdenfors
Cognitive Science, Lund University, Lund, Sweden
e-mail: peter.gardenfors@lucs.lu.se

M. Kaipainen et al. (eds.), *Conceptual Spaces: Elaborations and Applications*, Synthese Library 405, https://doi.org/10.1007/978-3-030-12800-5_1

self-standing, we have arranged them into four thematic sections. We now give a brief overview.

In *Conceptual Spaces, Generalisation Probabilities and Perceptual Categorisation*, Nina Laura Poth connects models of stimulus generalization from psychology with conceptual spaces. Starting with Shepard's law of universal generalization, she discusses Tenenbaum and Griffith's partial critique of Shepard's proposal. Poth's contribution consists in showing how to reconcile these two viewpoints by adopting the conceptual spaces framework. She particularly shows how the framework accounts for probability assignments, both statically and dynamically. Her central claim is that conceptual spaces help improve Bayesian models while offering an explanatory role for psychological similarity.

In *Formalized Conceptual Spaces with a Geometric Representation of Correlations*, Lucas Bechberger and Kai-Uwe Kuehnberger use conceptual spaces to bridge the gap between symbolic and subsymbolic AI by proposing an intermediate conceptual layer based on geometric representations. To answer how an (artificial) agent could come to learn meaningful regions in a conceptual space from unlabeled perceptual data (aka Harnad's symbol grounding problem for concept formation), the authors devise an incremental clustering algorithm. This groups a stream of unlabeled observations (represented as points in a conceptual space) into meaningful regions. If natural concepts are represented as convex regions of multidimensional conceptual spaces, however, this bars from representing correlations between different domains. Using a parametric definition of concepts, Bechberger and Kühnberger therefore propose a formalization based on fuzzy star-shaped sets. (Star-shapedness is a weaker constraint than convexity but still allows for prototypes.) They also define a comprehensive set of operations both for creating new concepts based on existing ones and for measuring relations between concepts.

In *Three Levels of Naturalistic Knowledge*, Andreas Stephens adopts an epistemological approach, and further refines the tripartite description of knowledge in Gärdenfors and Stephens (2018), who argued for three nested basic forms of knowledge: procedural knowledge-how, conceptual knowledge-what, and propositional knowledge-that. In this volume he investigates and integrates this description in terms of knowledge-accounts adopted from cognitive ethology and cognitive psychology. According to his chapter, semantic memory (which he interprets as conceptual knowledge-what) and the cognitive ability of categorization can be linked together. He argues that the conceptual spaces framework is particularly well suited to model this relation.

In his *Reply to José Hernández-Conde*, Peter Gärdenfors responds to Hernández-Conde's recent criticism of how to interpret the convexity criterion in the theory of conceptual spaces. Gärdenfors stresses that he devised the criterion as an empirically testable law for concepts, one that he intended as a necessary (rather than a sufficient) condition for a *natural* concept. Therefore, he submits, the range of "cases where the convexity criterion could be violated [...] shows that the criterion is rich in empirical content."

In *On the Essentially Dynamic Nature of Concepts: Constant if Incremental Motion in Conceptual Spaces*, Joel Parthemore claims that concepts are not only subject to change over an iterative lifecycle, but remain in a state of continuous motion. He offers three theses to the effect that concepts must change, must have a life-cycle, but must not change too much, either, in order to reach reasonable stability. The chapter's central claim is that concepts are in constant, if incremental motion. Each application of a concept, as it were, thus causes ripples throughout the conceptual system. Moreover, the change takes place at all the system's levels, including even mathematical concepts. One conclusion of the chapter is that, if concepts are in a state of continuous motion, then conceptual spaces must change and adapt, too.

In *Seeking for the Grasp: An Iterative Subdivision Model of Conceptualization*, Mauri Kaipainen and Antti Hautamäki investigate the speculative claim that the faculty to conceptualise may have developed with *homo habilis*'s ability to manage concrete actions in space and time. The authors thus propose an analogy between the cognitive grasping of concepts and the physical grasping of objects. On this basis, they offer a perspectival elaboration of conceptual spaces, which views "concepts as transient constructs that emerge from continuous dynamic cognition," thus elaborating the notion of perspective-dependent concepts by offering an iterative model.

In *Lost in Space and Time: A Quest for Conceptual Spaces in Physics*, Sylvia Wenmackers addresses whether all physical concepts are amenable to modelling in conceptual spaces, investigating whether dimensions in physics are analogous to quality dimensions such as the dimensions of colour space. The focal concepts are the domain of force in classical mechanics and the time dimension as it is used in classical and relativistic physics. Wenmackers obtains strong parallels between the dimensions of physics and those of conceptual space, but also finds that the development of physics has led to ever more abstract notions of space, for example phase spaces, that may not be translatable into the theory of conceptual spaces directly.

In *Interacting Conceptual Spaces I: Grammatical Composition of Concepts*, Josef Bolt, Bob Coecke, Fabrizio Romano Genovese, Martha Lewis, Dan Marsden and Robin Piedeleu present part of an ambitious project to combine descriptions of the structure of language in terms of category theory with semantic structures based on conceptual spaces. The central new idea in the paper is a construction of the category of convex relations that extends earlier category-theoretical descriptions. On the basis of conceptual spaces, the authors then provide semantic characterizations of nouns, adjectives, and verbs in terms of the category of convex relations. This shows how such characterizations generate a novel way of modelling compositions of meanings.

Finally, In *Magnitude and Number Sensitivity of the Approximate Number System in Conceptual Spaces*, Aleksander Gemel and Paula Quinon address the domain of numerical cognition, which studies correlations between number and magnitude sense. Their study applies the theory of conceptual spaces to characterize the Approximate Number System (ANS), the innate core cognitive systems proposed by Dehaene (1997/2011), and particularly models the quantitative representations that make for the system's fundamental assumptions.

References

Gärdenfors, P. (2000). *Conceptual Spaces: On the geometry of thought*. Cambridge, MA: The MIT Press.
Gärdenfors, P., & Stephens, A. (2018). *Induction and knowledge-what*. European Journal for Philosophy of Science, 8(3), 471–491.
Dehaene, S., & Cohen, L. (1997). *Cerebral pathways for calculation: Double dissociation between rote verbal and quantitative knowledge of arithmetic*. Cortex, 33(2), 219–250.
Dehaene, S. (2011). *The number sense: How the mind creates mathematics*. New York: Oxford University Press.

Part I
Concepts, Perception and Knowledge

Chapter 2
Conceptual Spaces, Generalisation Probabilities and Perceptual Categorisation

Nina L. Poth

Abstract Shepard's (Science 237(4820):1317–1323, 1987) *universal law of generalisation* (ULG) illustrates that an invariant gradient of generalisation across species and across stimuli conditions can be obtained by mapping the probability of a generalisation response onto the representations of similarity between individual stimuli. Tenenbaum and Griffiths (Behav Brain Sci 24:629–640, 2001) Bayesian account of generalisation expands ULG towards generalisation from multiple examples. Though the Bayesian model starts from Shepard's account it refrains from any commitment to the notion of psychological similarity to explain categorisation. This chapter presents the conceptual spaces theory as a mediator between Shepard's and Tenenbaum & Griffiths' conflicting views on the role of psychological similarity for a successful model of categorisation. It suggests that the conceptual spaces theory can help to improve the Bayesian model while finding an explanatory role for psychological similarity.

2.1 Introduction

As a counter to the behaviouristically inspired idea that generalisation of a particular kind of behaviour from one single stimulus to another single stimulus is a mere failure of discrimination, Shepard (1987) formulated a law that he empirically demonstrated to obtain across stimuli and species. His argument was that the law models categorisation as a cognitive function of perceived similarities.

The ULG has contributed to many models in categorisation research. One such a model that evolved on the basis of his work is Tenenbaum and Griffiths'(from herein T&G, 2001) Bayesian inference model of categorisation. T&G argue that

N. L. Poth (✉)
School of Philosophy, Psychology and Language Sciences, University of Edinburgh, Edinburgh, UK
e-mail: nina.poth@ed.ac.uk

M. Kaipainen et al. (eds.), *Conceptual Spaces: Elaborations and Applications*, Synthese Library 405, https://doi.org/10.1007/978-3-030-12800-5_2

their model is advantageous to Shepard's model in two respects. On the one hand, it can capture the influence of multiple examples on categorisation behaviour. On the other hand, T&G argue that it can unify two previously incompatible approaches to similarity. One is Shepard's approach to similarity as a function of continuous distance in multi-dimensional psychological space. The other is Tversky's (1977) set-theoretic model of similarity, which considers the similarity of two items to be a function of the number of their shared or distinct features. T&G argue that their model is advantageous to Shepard's original proposal because it is formally compatible with both conceptions of similarity and thus scores high in terms of its unificatory power. However, T&G take as an implication of the fact that their model is not strictly committed to any particular conception of similarity (i.e. Shepard's or Tversky's), that the (scientific) concept of similarity can be generally dismissed from explanations of the universal gradient of generalisation that Shepard had observed; probabilities alone are sufficient. Contra Shepard, T&G thus suggest considering generalisation probabilities as primary to similarity.

In this chapter, I suggest that the theory of conceptual spaces offers a perfect tool for resolving this debate. In particular, I argue that the theory of conceptual spaces can make T&G's Bayesian model more conceptually transparent and psychologically plausible by offering a tool to supplement it with a psychological similarity space, while capturing its advantage of showing that the multiplicity of examples in a learner's history matters for changes in categorisation behaviour. The conceptual spaces theory then helps to explicate that some notion of similarity is indeed needed for probabilistic models of categorisation more generally, and hence keeps in with Shepard's original motivation to explain categorisation as a function of perceptual similarity.

In Sect. 2.2, I outline Shepard's (1987) model of categorisation, with an emphasis on the role he attributes to perceived similarities in categorisation. In Sect. 2.3, I present T&G's (2001) expansion of Shepard's model, with an emphasis on the size principle – a principle which formally expresses the added value from considering multiple examples for categorisation. In Sect. 2.4, I present some problems with T&G's model. I argue that T&G's conclusion that probabilities should be considered primary to similarities is not warranted and that this perspective undermines their model's semantic interpretability. In Sect. 2.5 I suggest, more positively and on the basis of Decock and colleagues' (2016) Geometric Principle of Indifference, to consider a conceptual space as a semantic basis for Bayesian hypotheses spaces. I argue that in providing such a space, the conceptual spaces theory can help to avoid the issues with T&G's model and bring it in line with Shepard's original motivation to explain the generalisation gradient as a psychological function of similarity. I conclude with a suggestion, that this combination of a conceptual space and Bayesian inference could be considered as a more fruitful approach to modelling generalisation probabilities in perceptual categorisation than a probabilistic model on its own.

2.2 Shepard's Universal Law of Generalisation

This section briefly reviews Shepard's mathematical model of generalisation, and points out its historical relevancy for cognitive-representational models of categorisation.

Shepard (1987)'s universal law of generalisation is a pioneering concept in the psychology of perceptual categorisation. When Shepard published the results of his work, it was widely held that categorisation does not follow a universal law or pattern that might reflect a natural kind structure. Shepard contrasts the case of categorisation to Newton's (1687/1934) universal law of gravitation. Newton's law was very influential and helped to discover invariances in the physical structure of the universe. Inspired by Newton's achievements, Shepard's aim was to find a mathematical generalisation function that accurately models the psychological representation of cognitive categories by extracting the invariances in the perceived members of a category. Shepard took this to be a vital move against behaviourism and for the idea that generalisation is a cognitive decision, not merely a failure of sensory discrimination.

Shepard's law can be expressed by the following proposition.

(ULG) The universal law of generalisation:
For a set of stimuli, i and j, the empirical probability of an organism to generalise a type of behaviour towards j upon having observed i is a monotone and exponentially decreasing function of the distance between i and j in a continuous psychological similarity space.

ULG states that with a continuous increase in the distance between stimuli i and j in psychological space (that is, with an increase in their perceived dissimilarity), subjects are decreasingly likely to give these stimuli the same behavioural response. On this basis, ULG predicts that subjects should be less likely to generalise a behaviour associated with a given physical stimulus towards a relatively dissimilar stimulus and more likely to generalise the behaviour towards a relatively similar stimulus.

Shepard captured this tendency formally in terms of an exponential decay function. He obtained this function by plotting the probability of generalisation (the observed relative frequency at which a subject generalises behaviour to stimulus i towards stimulus j), g_{ij}, against a measure of psychological stimulus distance, d_{ij}, where psychological distance was obtained by means of the multi-dimensional scaling method (Carroll and Arabie 1980; Kruskal 1964; Shepard 1962). Shepard showed that the generalisation gradient is invariant across various stimuli (e.g. size, lightness & saturation, spectral hues, vowel and consonant phonemes and Morse code signals) and across species (e.g. pigeons and humans), thus the name 'universal law'. He obtained two insights from the mathematical modelling of this law.

1. If measured based on psychological (instead of physical) distance, the shape of the generalisation gradient is uniform across stimuli and species. (Uniformity)
2. The metric of the psychological similarity space is either the City-Block distance or the Euclidean metric. (L_1-/L_2-measurability)

The first point expresses the idea that differences in stimulus strength and corresponding generalisation might depend on differences in the psychophysical function that transforms physical measurements into psychological ones. For example, subjects might generalise the same label to two colour shades that a physicist would classify as 'green' and 'yellow' along the physical wavelength spectrum. However, if the two colour shades are positioned in a model of psychological colour space instead[1] then subjects' generalisations might be expected. This is because in psychological similarity space, the colour shades may be judged to be more similar than their measure along the physical wavelength spectrum actually indicates. The physical distance between stimuli along the one-dimensional wavelength spectrum might thus differ from their perceived distance in multi-dimensional psychological similarity space. Shepard took this discrepancy as a possible explanation of the previous difficulty to establish an empirically adequate model of generalisation by measuring physical stimulus space. Thus, a transformation function would be needed to recover a psychological distance measure from the physical distance measure. Shepard's second insight was that this function is a member of the family of Minkowski metric.

Categories in this psychological framework are modelled as consequential regions in multidimensional similarity space. Shepard assumes three constraints on the categoriser's background beliefs about consequential regions prior to any observation.

> ...(i) all locations are equally probable; (ii) the probability that the region has size s is given by a density function $p(s)$ with finite expectation μ; and (iii) the region is convex, of finite extension in all directions, and centrally symmetric. (Shepard 1987, 1320)

(i) is important because it yet does not assume that there are differences in the internal structure of categories. This is relevant because if any possible item in a category has the same chance of occurring, then this model cannot account for prototype effects in categorisation (cf. Rosch and Mervis 1975). (ii) is advantageous with respect to the formal precision and flexibility of the model. Since the magnitudes of the measured stimuli (e.g. brightness and sound) are measured in continuous space, probability densities are a suitable tool to use for a decision strategy when evaluating candidate categories on the basis of the training stimulus. (iii) is an assumption that makes the model mathematically more elegant, but Shepard has given additional arguments for assuming that categorisers indeed categorise in ways that satisfy convexity (Shepard 1981).

[1]A common example is the 3-dimensional CIELAB colour space which models colour representations along three axes, the hue, saturation and brightness dimensions (cf. Fairchild 2013).

As a psychophysical law, ULG explains the probability of generalisation by heavily relying on the notion of internally represented similarities. What makes the generalisation function psychological as opposed to physical is that "it can be determined in the absence of any physical measurements on the stimuli." (Shepard 1987, 1318). For example, even if the colour of an item on the physical wavelength spectrum might change so as to become more different under differing lighting conditions, this change might not actually be represented as a change in the vector coordinates that would be assigned to the perceived colour of the item in psychological similarity space. Thus, the invariance of the gradient could not be explained without the more subjective notion of representational perceptual similarity.

If this is correct, and the generalisation gradient arises from psychological instead of physical measurements, then what is needed for a model of similarity-based categorisation is a conceptual distinction between the psychological and physical magnitude of the difference between (training- and test-) stimuli, respectively, how they are related. In his work on psychophysical complementarity, Shepard (1981) argues that psychological similarity offers such a distinction. In particular, Shepard distinguishes between two kinds of similarities; first- and second-order similarities. Accordingly, first-order similarities are similarities between physically measurable properties in the world on the one hand and representations thereof on the other hand. For example, consider the similarity between the redness of a dress as measured on the physical wavelength spectrum, and the redness of the dress as I perceive it. Second-order similarities, in contrast, are similarities between mental representations themselves. For instance, consider the similarity between my representation of the dress' redness at one point in time, t_0, as compared to my perceptual experience of the dress' redness at another point in time, t_1.

Why should it be important to distinguish between first- and second-order similarities for categorisation? Because they impose different kinds of accuracy conditions on the representations of similarities, which in turn constitute the generalisation gradient. Edelman (1998) motivates this point by alluding to Shepard & Chipman's (1970) distinction between first- and second-order isomorphisms.[2] They suggest that veridicality in perception is instantiated through the perception of similarities amongst the structure of shapes. The task of perceptual categorisation is not to build representations that resemble objects in the world. Instead, the task of the visual system is to build representations that stand in some orderly relationship to similarity relations between *perceived* objects. This supports Shepard's idea that the criteria for whether generalisation is accurate or not are not determined by physical measurements but by some psychological standard.

[2]For instance, second-order isomorphisms of shapes are measurements of similarities between representations of the similarities of shapes (Edelman refers to this as a 'representation of similarity'), as opposed to similarities between distal shapes and their proximal representations (a 'representation by similarity'). Where categories are seen here as reference shapes, they are to some extent cognitive constructions.

Consider second-order isomorphisms between shapes. What is represented are principled quantifications of changes of shape, not shapes themselves. The idea is that information about distal (represented) similarity relationships is picked up by the representing system through a translation process. In particular, the information about similarity relationships is reduced to invariances between movements in distal parameter space. This allows for dimensionality reduction to constitute a proximal (representing) shape space (Edelman 1998, 452). This process of translation (as opposed to a process of reconstruction), allows for a reverse inference from subjects' similarity judgements to a common metric. Veridicality, in that sense, means consistency amongst subjects when judging the similarities between considered object shapes, as opposed to consistency across stimuli conditions (for individual shapes).

The example of first- and second-order similarities illustrates that Shepard considers psychological similarity as explanatorily central to the relationship between assignments of category membership and perceived similarity. This goes against behaviourist analyses because similarity is seen as a cognitive function of decreasing distance. But Shepard's model is restricted to a comparison between representations of single members of a category. An alternative view on generalisation probabilities is offered by a Bayesian model of categorisation that considers generalisation from multiple examples but eventually suggests to explain categorisation regardless of the notion of psychological similarity. The model is presented in the next section.

2.3 Tenenbaum and Griffiths' Size Principle

This section outlines a Bayesian model of categorisation by Tenenbaum and Griffiths (2001) that attempts to expand Shepard's approach to generalisation in two ways.

1. They show that the number and magnitude of examples observed shapes the generalisation gradient.
2. They show that the probability of generalisation is (formally) independent of any particular model of similarity.

I elaborate shortly on both points to illustrate the differences between T&G's and Shepard's views on the relation between generalisation probabilities and psychological similarities.

The first point of expansion considers the generalisation function that learners are supposed to follow when learning categories. For this, T&G suggest a Bayesian inference algorithm, which they call the *size principle*. It helps to consider the size principle in light of the general Bayesian learning theory that T&G suggest.

The idea is that learners follow Bayes' theorem in computing the posterior probability, $Pr(H|E)$, of a hypothesis, H, about which consequential region is shared for stimuli of a common class, in light of the available evidence, E.

Bayes' Theorem

$$Pr(H|E) = \frac{Pr(H)Pr(E|H)}{Pr(E)}$$

Bayes' Theorem makes explicit how the posterior probability of a hypothesis given some piece of evidence can be obtained; by taking the prior probability, $Pr(H)$, together with the likelihood, $Pr(E|H)$, relative to the probability of the evidence, $Pr(E)$. For the current purpose, only the prior and likelihood are of interest. This is because dividing by $Pr(E)$ only serves normalisation purposes.

T&G argue that the likelihood term can be replaced by the size principle. The size principle states that if the available evidence is held constant, hypotheses that point towards smaller consequential regions should be preferred over those hypotheses that suggest larger consequential regions when making a generalisation decision. Moreover, if the information about perceived similarities is held constant, the tendency to prefer smaller categories for generalisation should become stronger with an increasing number of examples observed for that category. Formally, the size principle can be expressed as follows.

The size principle

$$Pr(E|H) \propto \left(\frac{1}{size(H_C)}\right)^{|n|} \tag{2.1}$$

Consider an example for the size principle that comes from Xu and Tenenbaum (2007). Three Dalmatians are given as examples for the word 'fep', together with the following hypotheses space.

H_1: $\langle$'fep', DALMATIAN$\rangle$
H_2: $\langle$'fep', DOG$\rangle$
H_3: $\langle$'fep', WHITE WITH BLACK DOTS$\rangle$

If the three Dalmatians are a random sample of the true category that the word 'fep' refers to, the size principle says that learners should have a higher degree of belief in H_1 than in H_2 and H_3. Following this size principle, this is a rational choice because it is more likely to observe 3 Dalmatians as examples of what 'fep' means if in fact it referred to the category of Dalmatians as compared to the class of dogs or things that are white and have black dots. The size principle thereby expresses what Xu and Tenenbaum (2007) call a *suspicious coincidence* mechanism: it would be very unlikely to observe 3 Dalmatians if 'fep' meant 'dog'. More formally, based on the size principle: since $size(Dalmatian) < size(dog) < size(white\ with\ black\ dots)$, $Pr(E|H_1) > Pr(E|H_2) > Pr(E|H_3)$.

The second point of T&G's expansion takes the size principle for granted and argues that the probability of generalisation – expressed in the likelihood term – is itself primary to perceived similarities in the analysis of categorisation behaviour.

This claim is motivated by the idea that formally, the probabilistic model can be translated into either a set-theoretic approach to similarity (such as the one put forward by Tversky 1977) or a similarity-as-psychological-distance approach (such as the one suggested by Shepard 1981, 1987). The probability space that is considered in T&G's model can be measured in terms of both, the number of shared or distinct features or the relative distances between items in a continuous space. This is because relative frequencies can be expressed in terms of either spatial or numeral proportions. Thus, based on which quantity the size of a consequential region is actually measured is irrelevant to Eq. 2.1.

T&G take the formal independence of probabilities from the particular type of similarity model to indicate that perceptual similarities might be derived from generalisation probabilities. T&G conclude that generalisation probabilities are enough to explain categorisation behaviour and that such explanations can therefore do without the concept of psychological similarity. They write:

> We expect that, depending on the context of judgement, the similarity of y to x may involve the probability of generalizing from x to y, or from y to x or some combination of those two. It may also depend on other factors altogether. Qualifications aside, interesting consequences nonetheless follow just from the hypothesis that similarity somehow depends on generalization, without specifying the exact nature of the dependence. (Tenenbaum and Griffiths 2001, 637)

However, it is yet an open question how probability assignments can determine perceived similarities between category members. More on this problem in Sect. 2.4.

2.3.1 *Advantages Over Shepard's Model*

The size principle adds to Shepard's model in two ways. One advantage is that T&G's model makes the role of exemplar variability for generalisation more transparent. This helps understanding why multiple examples, expressed by the exponent $|n|$, can help to make categorisation more precise. Accordingly, a higher number of examples helps to gain more information about where to set the boundaries of a category: "All other things being equal, the more examples observed within a given range, the lower the probability of generalization outside that range." (Tenenbaum and Griffiths 2001, 633). Categorisation is then a form of prediction. A category plays the role of a random variable, $X = \{x_1, \ldots, x_n\}$ whose probability to take on particular values in similarity space must be calculated. A more informative distribution of the estimated random variable allows for more precise predictions which values in similarity space are most probably to be observed as next examples of the to-be-inferred category.

Another advantage of T&G's model over Shepard's model is that it is richer in the predictions that it makes about generalisation performance. This also follows from the additional consideration of exemplar variability in the size principle. If less data is available under equal exemplar variability, the uncertainty in generalisation is greater and the overall distribution of probability assignments less distinct. Holding the exponent stable, the size principle prescribes a preference for categories with a smaller magnitude. Exemplar-variability, accordingly, determines generalisation probability. To that end, T&G can explain changes in the generalisation gradient that occur despite constancies in similarity comparisons. For instance, the model predicts that even under circumstances in which a novel item is actually very similar to the average of previously observed exemplars, it might not be generalised upon. The rationale is given by the size principle. With an increasing amount of examples, rational learners should become more restrictive in their willingness to expand generalisation. In other words, boundaries around regions in similarity space become sharper and generalisation patterns more distinct throughout a subject's learning history. Thereby, the Bayesian model can account for undergeneralisation – some of the presented examples are not considered particularly relevant for generalisation, even though they would fall within the hypothesised consequential region. Since this result is difficult to obtain in Shepard's model, the size principle offers a valuable expansion of ULG.

2.4 Problems for Tenenbaum and Griffiths' Approach

This section outlines three problems for T&G's approach. The first problem is that T&G are too hasty in dismissing similarity from their programme of explaining generalisation. T&G illustrate that their Bayesian model is independent of any particular account of similarity. However, T&G conclude that just because the Bayesian model is formally independent of any particular notion of similarity, similarity can be dismissed from an explanation of the generalisation gradient. They even make the stronger claim that the represented generalisation probabilities should be seen as primary to the perceived similarities of category members. But this conclusion seems too hasty. Just because their model is independent of any particular view on similarity (i.e. it is formally compatible with both Tversky's and Shepard's definition of similarity), this does not mean that similarity more generally cannot serve as a conceptually useful notion for the probabilistic model. Indeed, it could be plausible to think that considerations of similarity would, in fact, make T&G's Bayesian inference approach more precise and explanatorily useful. The first problem has two facets in effect.

The first facet of the problem of disregarding similarity is that similarity is conducive to an explication of the relationship between exemplar variability and how this changes the generalisation gradient. The reason for this is that it is hard to see how the concept of exemplar variability itself can be defined without alluding to similarity in the input data; in fact, it is easier to see how generalisation probabilities can be derived from changes in the average similarity between examples. This also corresponds to the common conception of exemplar variability in the literature on clustering algorithms in machine learning, particularly on nearest-neighbour models (Russell and Norvig 2002, ch. 18). The remaining question for T&G's model is thus which of its features can establish the relationship between the observed generalisation gradient and the agent's corresponding subjective degrees of belief about category membership, if not similarity.

The second facet is that T&G underestimate the semantic value of similarity for interpreting the size principle. Yet Eq. 2.1 leaves open how the size of a candidate category should be measured, and its semantic content interpreted. One reason for its semantic opaqueness is that Eq. 2.1 loses the explicit mentioning of the evidence on the right-hand side of the equation.[3] This also makes it difficult to identify the degree of confirmation of a belief by an example because it cannot be compared how far and by which measure the contents of terms E and H align. More conceptually, it is not clear what the belief in the Bayesian model that is assigned a probability value is actually about. Similarity could provide an answer to this question. For instance, on Shepard's account, consequential regions are individuated through a measure of distance. This measure is an internal representation of the agent, and thus provides content for a belief about category membership. Similarity can help understanding of the relationship between the perceptual features of observations as represented by the agent and the probabilistic inference of concepts. Thus, T&G should not disregard but instead consider similarity as a conceptual basis for their Bayesian categorisation model.

The second problem with T&G's Bayesian approach is that it is incomplete. This is because it only considers the likelihood and leaves out considerations of prior constraints on the inference of categories. For instance, preferences for some over other categories might be important for deciding about cases in which the observed evidence confirms multiple candidate categories equally well.

A final problem with T&G's approach is that the size principle invites worries about undergeneralisation. Undergeneralisation occurs when a to-be-learned word is applied to only a subclass of the items that denote its true meaning. For example, if the word 'dax' meant 'tulip', but a learner applies it only to yellow tulips, then 'dax' is undergeneralised. In Eq. 2.1, undergeneralisation is to be expected because a learner should prefer to generalise towards the smallest possible category that is compatible with the evidence. This is problematic for the empirical adequacy of a categorisation model because we know that category learners, such as children, do not only undergeneralise but often do overgeneralise word meanings to broader categories than would be accurate (cf. Bloom 2002, 36, 158). It is not clear how T&G's model can capture an optimal trade-off between under- and overgeneralisation.

[3] I thank Wolfgang Schwarz for pointing me towards this issue.

2.5 A Geometric Principle of Categorisation

This section uses the conceptual spaces theory to move beyond the size principle. In particular, the argument is that the size principle can be made more precise by considering a conceptual space as a semantic basis for Bayesian inference. I start by outlining Decock et al.'s (2016) argument that the conceptual spaces theory can be used to formulate a geometric principle of indifference as a solution to the different carvings-up problem. I subsequently argue that Decock et al.'s geometric solution to the different carvings-up problem can be generalised towards a solution to the problems with T&G's approach. This generalisation is framed as a geometric size principle that moves beyond T&G's size principle in making it more precise. Though being motivated by a similar idea, the geometric size principle also moves beyond Decock et al.'s approach in that it can account for category learning.

2.5.1 Decock et al.'s Geometric Principle of Indifference

Decock et al. (2016) use a conceptual space to establish a geometric principle of indifference (gPOI), and thereby avoid some of the problems that were encountered by a standard principle of indifference (sPOI). The sPOI states that

> ...given a set of mutually exclusive (at most one can be true) and jointly exhaustive (at least one must be true) propositions, and barring countervailing considerations, one ought to invest the same confidence in each of the propositions. Put differently, given a set of propositions of the aforementioned kind, if you lack any reason *not* to treat them evenhandedly, you *should* treat them evenhandedly. (Decock et al. 2016, 55)

A problem for the sPOI is the different carvings-up problem. This is the problem of choosing the right kind of hypotheses space. For example, considering a box with an unknown number of multi-coloured marbles, one can carve up the space of considered possibilities for any outcome in different ways. Probabilities could be distributed over the set of hypotheses 'red' and 'any other colour'. Or, alternatively, one could distribute probabilities over the set of hypotheses, 'red', 'blue' or 'any other colour'. The first option can be expressed as $\mathscr{H} = \{H_1, H_2\}$, where H_1 ='red' and H_2 = 'any other colour'. The second option can be expressed as $\mathscr{H} = \{H_1, H_3, H_2\}$, where H_1 = 'red', H_3 = 'blue', and H_2 = 'any other colour'. Which set of hypotheses is considered matters for the distribution of the probability assignments because these are mutually dependent.

The problem is that these different ways to carve up the space are each probabilistically admissible but taken together, they become incoherent. Considering the first hypotheses space, one could assign $Pr(H_1) = .6$ and $Pr(H_2) = .4$. But considering the second hypotheses space one could assign $Pr(H_1) = .25$, $Pr(H_2) = .5$ and $Pr(H_3) = .25$. How can it be rational to assign $Pr(H_1) = .6$ and $Pr(H_1) = .25$ simultaneously? Intuitively, the same probability value should be assigned to H_1 in each case.

Decock et al. solve the different carvings-up problem in two steps. In a first step, they regard the geometry of concepts as primary to the formulation of the degrees-of-belief functions. For this, they use the architecture of concepts as suggested by the conceptual spaces theory (Gärdenfors 2000).

A conceptual space is a geometric similarity space. It is defined as a number of quality dimensions (e.g. height, size, hue, saturation, brightness). Objects can be assigned a value along each dimension, where each value represents the respective perceived quality of the object. In combining those values from each axis of the conceptual space, objects can be represented as vectors in conceptual space. The distances between the vectors express the perceived similarity between them. Roughly, the larger the distance, the less similar the objects are.[4] Regions in conceptual space represent concepts – cognitive categories – and cover areas in conceptual space. The content of a region captures not only information about already observed members of a category, but also information about yet unobserved members. Thus, a region in conceptual space indicates a concept's intensional content, where the intension could be understood as all the possible qualities that a member of the concept can be assigned (cf. Carnap 1988).[5]

On Decock et al.'s account, the basic quality dimensions are important for specifying the gPOI. This is because they are taken as the fundamental attributes (e.g. attributes such as *shape, size, colour*) needed to define the predicates used in the formulation of the hypotheses.

In a second step, Decock et al. use a modification of Carnap's (1980) γ-rule to specify the prior probabilities for any possible carving up. The γ-rule specifies how the probability of an object, o's (e.g. an unobserved colour shade), falling in a region, C_i, can be computed. The $gamma$-rule says that this probability is equal to the size of C_i relative to the size of the conceptual space, CS (which in Carnap's terminology is the attribute space). Where the sentence 'an object o's falling in a region C_i' refers to the content of a hypothesis, $H_{C_i} : \{o \in \text{region}_{C_i}\}$, the rule can be expressed more formally as follows.

[4]To be a bit more precise, the distance function is exponential. Gärdenfors (2000) suggests the Euclidean distance metric for Euclidean space and the Minkowskian distance metric for non-Euclidean space.

[5]Categorisation in conceptual spaces follows a function that maps each point in similarity space onto a unique cell in a Voronoi tessellation. The fundamental categories that serve as candidates when formulating the hypotheses prior to their evaluation thus result from a mechanism of concept acquisition which requires justification itself. For the purpose of this chapter, I mainly disregard the Voronoi tessellation as a concept acquisition mechanism, and consider only the geometric properties of an established conceptual space as substantial for the argument.

Carnap's γ-rule

$$Pr(H_{C_i}) = \frac{size(region_{C_i})}{size(CS)} \tag{2.2}$$

Decock et al. (2016) argue that Eq. 2.2 can solve the different carvings-up problem because it helps to fix the probabilities for each considered hypothesis (e.g. H_{C_i} : $\{o \in \text{region}_{C_i}\}$ versus H_{C_j} : $\{o \in \text{region}_{C_j}\}$) in dependence of the size of the underlying region (e.g. region_{C_i} versus region_{C_j}). Thus, considering two possible hypotheses spaces, it can be exactly decided how the space should be carved up because the concepts used for formulating the relevant propositions are now fixed in their relative sizes.

Consider the following example. Take C_1 to stand for the region in conceptual space representing the category *red*. Take C_2 to stand for the region in conceptual space representing the category *blue*. And take C_3 to stand for the region in conceptual space representing the category *any other colour*. Following Eq. 2.2, the prior probability of a hypothesis pointing towards an object o to lie in C_1 is the area of C_1 divided by the area of the entire conceptual space. Decock et al. (2016) call this prior probability measure α. Then, it does not matter how the space is carved up, that is, whether the partition $\{(o \in C_1), (o \in C_2)\}$ or the partition $\{(o \in C_1), (o \in C_2), (o \in C_3)\}$ is considered. The assignment of a prior degree of belief to $(o \in C_1)$ will be the same, namely α, regardless of whether any other colour is considered in the prediction. With a similar example, Decock et al. show that by fixing probabilities based on the spatial properties of the conceptual space, it is possible to stay neutral on which partition is the right one – the prior degrees of belief stay the same.

Decock et al.'s approach illustrates a further advantage of the conceptual spaces theory. This is that the geometric space can help making the γ-rule more precise. In particular, Decock et al. use the geometric properties of the conceptual space to establish a unique measure of the size of a region – the Lebesgue-measure. Thereby, they can make more explicit the relationship between the assignments of probabilities to candidate hypotheses (i.e. beliefs) in dependence of their semantic contents (i.e. areas in conceptual space). More formally, their modification looks as follows.

Decock et al.'s Lebesgue-specification of the γ-rule

$$\mu^*(C_i) = \frac{\mu(C_i)}{\mu(CS)} \tag{2.3}$$

For Decock et al., the normalised Lebesgue measure, μ^*, represents the prior degree of belief of the agent to classify an unknown object as a member of concept C_i. Thus, Eq. 2.3 makes Eq. 2.2 more precise. This is important because just upon considering the conceptual space, it is possible to semantically interpret the hypotheses. This is because the conceptual space provides the basic predicates without which the hypotheses could not be formulated. The specification is thus possible only via the assumption of the conceptual space. In particular, the *size* is measured by integrating over the subset of vectors in conceptual space that would be covered by each category predicate, C_i. This way, the gPOI delivers a formally precise solution to the different carvings-up problem, and presents a better alternative to the sPOI.

More generally, Decock et al.'s account is similar in its spirit to the criticism of T&G's approach presented here. Roughly, just like the sPOI lacks a semantic basis for prior degrees-of-belief functions, T&G's size principle lacks a semantic basis for determining the likelihoods in Bayes' Theorem. The common claim is that the conceptual spaces theory offers a way out of the problem by making Bayesian inference more precise, and semantically interpretable.

2.5.2 Going Beyond the Size Principle

To avoid the problems with the size principle, I follow Decock et al.'s approach in two steps. The first step is to define the hypotheses space with the conceptual spaces theory. The second step is to explore in which ways the architecture of a conceptual space can provide additional constraints on categorisation.

For the first step, the conceptual spaces theory suggests that a category should be interpreted as a region in the conceptual space. It can be seen that under this interpretation, the size principle can be compared to Carnap's γ-rule. Equation 2.2 takes the size of the relevant region relative to the size of the entire conceptual space. In contrast, Eq. 2.1, disregarding for a moment the $|n|$-component, reverses this relation. An assignment of total probability (which represents that the observed item belongs to any category) is taken relative to the size of a candidate category. Thus, the size principle reverses the influence of the size of a category on the degree of belief function. Given the structural similarities between Eqs. 2.2, 2.3 and 2.1, also the size principle can be specified with the help of the geometry of concepts. Following Decock et al., the surface area of a region in conceptual space can be measured with the Lebesgue measure, μ (see Eq. 2.3). One way in which this could be expressed is as follows.

$$Pr(e_i|H_{\langle C_i,L\rangle}) = \frac{\mu(e_i \cap C_i)}{\mu(C_i)} \tag{2.4}$$

Equation 2.4 expresses the likelihood of a known item, e_i, given that it is a member of a region, C_i, in conceptual space. The likelihood is a measure of the relative overlap of a piece of evidence, $e_i \in E$, and a candidate region, C_i. This is taken relative to the measure of the region, C_i.[6] The likelihood hence becomes larger the better the relative overlap of the evidence with the candidate concept.

Equation 2.4 moves beyond the size principle by answering the problems that were associated with T&G's approach. It can solve the first problem of disregarding similarity too hastily because in contrast to T&G's argumentation, it does welcome similarity as a basis for probabilistic inference in categorisation. This is positive for avoiding the two facets that come with this problem. For the first facet, the geometric approach can offer at least an explanation sketch of how perceived similarities in varying examples relate to the corresponding probability assignments to possible future exemplars. In a conceptual space, exemplar variability is captured by measuring the average distance between the exemplars. A probability density function can be approximated once sufficiently many exemplars have been observed. In a conceptual space with continuous quality dimensions, the probability of observing an item from a particular region in conceptual space is then simply the area under the probability density function bounded by the region. A relation between the probability and the perceived similarities of the exemplars can thus be established based on the distance measure that a conceptual space is equipped with. In effect, this can give rise to the dynamics between the size of a category and how it determines the likelihood for the corresponding hypothesis (cf. Krumhansl 1978).

By considering similarity as a basis for probability assignments, the geometric approach also helps to find a more transparent semantic interpretation of the size principle. This is based on two changes that come with a geometric approach to the size principle. First, Eq. 2.4 replaces the likelihood with a measure of the relative proportion of areas in conceptual space. This defines the content of the hypotheses more precisely: as mappings from measurable regions in conceptual space (e.g. the concept TULIP) to the labels whose meaning is to be inferred (e.g. 'dax'). Second, Eq. 2.4 lets the term for the evidence, E, re-occur on the right-hand side of the equation. In doing so, the geometric approach helps to make more explicit the relationship between the evidence and a hypothesised category (here measured by their relative spatial overlap). Based on these two changes, Eq. 2.4 is easier to interpret than Eq. 2.1. The evidence is a point in conceptual space and a candidate category is a region in such a space, and their relative overlap can be measured geometrically. More generally, this shows that the geometric approach offers a more transparent way to interpret the likelihood term in the Bayesian inference of categories.

To make this point more clear, consider the following example. Zoey's dad wants to teach her the names of some colour categories. He shows her three particular colour shades, $blue_1$, $blue_2$ and $blue_3$, and he calls each of them 'azure'. For an overview:

$E = e_1 : \langle blue_1, \text{'azure'}\rangle, e_1 : \langle blue_2, \text{'azure'}\rangle e_3 : \langle blue_3, \text{'azure'}\rangle$

[6]This formal solution – to consider the relative overlap of the evidence and the candidate category – has already been suggested in a joint talk held by Peter Brössel and me at a conference in Salzburg, 2015 (Poth and Brössel 2015).

$$\mathscr{H} = H_1 : \langle turquoise, \text{`azure'}\rangle, H_2 : \langle light\ blue, \text{`azure'}\rangle, H_3 : \langle blue, \text{`azure'}\rangle$$

With the help of the conceptual spaces theory, a hypothesis can now be interpreted in terms of the average spatial distance between those subsets in conceptual space that are indicated by the corresponding regions. For instance, the content of H_1 can be partially expressed in terms of the distance of those vectors in colour space that would be perceived as turquoise colour shades. The other part of this content is the linguistic description itself, 'azure', which is given in the supervised learning environment that is considered here. In light of this interpretation, hypotheses can be evaluated in terms of the intensions of the categories that they point towards, that is, in dependence of their average spatial distances. Following Eq. 2.4, the information about perceived similarities as represented in conceptual space can then be used to assign probabilities to each candidate hypothesis. For instance, the semantic content of the hypothesis that $blue_1$, $blue_2$ and $blue_3$ belong to category *turquoise* is the relative overlap of the distance between $blue_1$, $blue_2$ and $blue_3$ and the average distance of perceivable items within the candidate region in conceptual space that represents the intension of the corresponding term (e.g. 'azure').[7] Then, $Pr(E|H_1)$ would win in the competition because it presents the best overlap with E.

Another way in which the geometric approach moves beyond the size principle is that it can make T&G's Bayesian inference model more complete. This is because whereas T&G yet only specify the likelihoods in the inference of categories, a conceptual space can offer additional constraints to determine the prior probabilities. One constraint is convexity. In the conceptual spaces theory, convexity means that if two items in conceptual space are known to belong to the same category, then any other item that would lie on a straight line between the two examples will be known to be a member of the category, too. Convexity can offer an additional constraint on word-meaning inferences in the context of categorisation tasks. This can be helpful for evaluating contexts in which the evidence is insufficient to decide for a unique way to generalise, e.g. when candidate categories are confirmed equally well by the evidence alone.

The motivation to build a convexity constraint in the architecture of a categorisation model comes from the idea that convex categories should be preferred in meaning inferences because they are easiest to infer. In the conceptual spaces theory, natural properties are considered to be those that are convex. This statement is expressed as "Criterion P" in the conceptual spaces theory (Gärdenfors 2000, 71). The rationale behind Criterion P is that it serves to establish a cognitive distinction between natural and gerrymandered categories. Such a distinction is needed to tackle Goodman's (1972) new riddle of induction: why do we want to prefer inferences towards categories such as *green* or *blue*, as opposed to those

[7]The k-nearest neighbour rule (Russell and Norvig 2002, ch. 18) suggests a similar interpretation. Here, to-be-classified items are grouped based on their distance to an average observation in a vector space, and this can be used to determine a model's graded meaning inferences.

towards *grue* or *bleen*? For Goodman, the answer is that *green* and *blue* are projectible predicates and should thus be preferred over *grue* or *bleen*, which are inductively useless. In the conceptual spaces theory, the conceptually abstract notion of projectibility is made more precise with the more spatially grounded notion of convexity.

To see how convexity can help narrowing down the candidate categories if the evidence alone cannot do so, imagine the following example. Over a period of three days, Zoey's dad shows her a sequence of three flowers. The flowers are of different colours but all have the same shape. On each of these observations, Zoey's dad says 'dax'. Thus, there is $E = e_1 : \langle$ white tulip, 'dax' $\rangle$, $e_2 : \langle$ yellow tulip, 'dax' $\rangle$, $e_3 : \langle$ red tulip, 'dax' $\rangle$. What is more, all observations are made during the afternoon. Given these three pieces of evidence, the following hypotheses seem plausible. $\mathscr{H} = H_1$: 'dax' means *tulip*, H_2 : 'dax' means *flower*, H_3: 'dax' means *tulip in the afternoon*.

Intuitively, H_1 is to be preferred over H_2 because H_1 is more plausible in light of E. Following Eq. 2.4, H_1 is better supported by E than H_2 because E and the candidate category *tulip* achieve a better relative overlap in conceptual space than E and the candidate category *flower*. However, E is insufficient to give H_1 an advantage over H_3 because all observations were in fact made in the afternoon. But intuitively, H_3 also appears less plausible than H_1. Equation 2.4 cannot provide sufficient reason for this because it only considers the role of the evidence in determining the probabilities. Thus, some other criterion must be chosen to evaluate the hypotheses in addition to the available evidence.

The convexity constraint can help here. The category *tulip in the afternoon* is most probably non-convex because it mixes the category *tulip* with the time dimension. Whereas objects that look like tulips are likely to form convex clusters in the shape and colour domains, adding the time dimension would make them become scattered in conceptual space. This is because by adding the constraint *in the afternoon*, objects that would be classified as *tulip* would lose the property of being a tulip during some intervals along the time dimension (e.g. during those intervals that represent the morning and the evening).[8] Under the assumption that *tulip in the afternoon* is non-convex, H_3, as opposed to H_1, points towards a gerrymandered category and should thus be harder to infer. In other words, H_1 should be preferred over H_3 because it points towards a more natural category, even if the evidence confirms both hypotheses equally well.

Formally, this suggestion will be best captured in the prior probabilities. That is, before Zoey is going to observe the next example, her prior degree of belief in the

[8]See also Gärdenfors' (2000, pp. 211) solution to the grue- and bleen problem, in which he argues that *grue* and *bleen* are non-convex because they are defined by the hue and the time dimensions. He argues that in contrast, the categories *blue* and *green* are convex because in the latter cases there is no interference with the time dimension.

hypothesis that 'dax' means *tulip* should be higher than her prior degree of belief in the hypothesis that 'dax' means *tulip in the afternoon*, just because *tulip* is a more natural concept.[9]

Apart from the specification of the prior probabilities, criterion P could also help with the worry of undergeneralisation. This is because if a category is convex, then it will have to occupy on average a larger proportion of the conceptual space than a category that maximally fits the examples. Though a full reply to this problem cannot be fleshed out here, one option is to consider the Voronoi tessellation (Gärdenfors 2000, pp. 87) to accommodate overgeneralisation. A Voronoi tessellation partitions the conceptual space into clusters of mutually exclusive and exhaustive convex sets of objects. The tessellation process starts from a prototype. At least two prototypes are needed to achieve a tessellation because the cells are established through a connection of the bisectors of hypothetical lines that connect the prototypes for any cluster of items. Upon tessellating the space, any item that is closer in space to a prototype than to any other prototype will be assigned a membership function for the category that is represented by the prototype, i.e. the cell associated with that prototype. With this method, a learner would overgeneralise easily, if the partition is broad enough. Generally, the approach outlined here should not be committed to the Voronoi tessellation. But it illustrates that convexity can indeed help to counter the worry of undergenralisation that comes with the size principle.

To recapitulate, this section has taken Decock et al.'s solution to the different carvings-up problem as starting point to help moving beyond the size principle. The advantage of the resulting geometric approach to categorisation is that it can avoid the problems presented in Sect. 2.4 while keeping in with Shepard's original motivation to use similarity as a tool to explain the generalisation gradient.

2.5.3 A Worry for a Geometric Principle of Indifference

Decock et al.'s approach has a problem; it cannot explain how information about how a category is commonly used changes the degree of belief in that category being the right meaning candidate. That is, it fails at accounting for pragmatic effects in category learning.[10] For instance, consider two categories, C_a and C_b, represented as regions in conceptual space. Imagine that the size of both corresponding regions is the same, as measured by the Lebesgue integral. That is, $size(C_a)/size(CS) = size(C_b)/size(CS)$. Because according to Carnap's γ-rule, all that matters for determining the prior degree of belief is the relative size of a category, the prior probabilities for C_a and for C_b must also be the same. Based on Decock et al.'s

[9]There is some evidence for a preference of convex categories in children learning homophones (e.g. Dautriche et al. 2016) and also amongst adults and other word types more generally Jäger (2010).

[10]Thanks to Peter Gärdenfors for pointing out this problem to me.

approach, we obtain that $\mu^*(C_a) = \mu^*(C_b)$ and thus that for any unknown item o, $Pr(o \in C_a) = Pr(o \in C_b)$. But imagine that it is also known that C_a is used often in linguistic communication, whereas C_b is very rarely used. It is not clear how this information could be captured by a measure of the size of either C_a or C_b alone. Thus, prior probabilities alone cannot capture the pragmatic influences on degrees of beliefs about category membership for unknown objects.

More generally, the problem is that Decock et al.'s approach cannot account for category learning. This is because it is limited to a static account of rationality, that is, one in which an agent is rational at a given point in time if and only if her degrees of belief can be represented by a probability function at that point in time. However, learning requires beliefs to change in response to novel evidence. Thus, for a rational-learning account of categorisation, a dynamic account of rationality that considers the evidence is needed. On such an account, an agent is rational if and only if her change in a degree of belief from an earlier to a later time point can be represented by conditionalisation.[11] That is, her degree of belief in H conditional on E prior to E's occurrence, $Pr(H|E)$, must equal her degree of belief in H after having learned E, $P_E(H)$. But to compute $Pr(H|E)$ and consider what happens upon learning E, a rational learner must follow Bayes' Theorem and compute the likelihood.

My suggestion is that information about how frequently a category is used is part of the evidence for category learning, and that such information can be captured in the likelihood term. Then, the geometric size principle would already contain the ingredients to accommodate the pragmatic challenge because it specifies the likelihood term.

A possible response from the geometric size principle could be made by stressing two points. First, the geometric approach specifies the likelihood instead of the prior probabilities. Second, it can also capture the role of multiple examples for category learning. Based on their differences, the geometric size principle is promising in accommodating for pragmatic effects in categorisation.

For the first point, I suggest that Decock et al.'s specification of Carnap's γ-rule and the modified size principle function as complementary elements in a Bayesian model of category learning. Formally, both equations express two different kinds of degrees of beliefs. Whereas the γ-rule replaces the prior probability of observing an unknown member, o, of a candidate category, $Pr(o \in C_i)$, the geometric size principle replaces the likelihood of observing an actual object, e_i, given that it is a member of a particular candidate category, $Pr(e_i|e_i \in C_i)$, where o and e_i can occupy the same point $\langle x, y, z\rangle$ in conceptual space.

For the second point, one could modify Eq. 2.4 to express the idea that the likelihood of observing a sequence of examples for a labelled category becomes proportionally greater the better the relative overlap of the sequence of examples is, on average, with the candidate region in conceptual space.

[11] The best argument for conditionalisation are diachronic Dutch Books (cf. Teller 1973).

$$Pr(E = \langle e_i, \ldots, e_n \rangle | E \in C_i) \propto \left(\frac{\mu^*(\langle e_i, \ldots, e_n \rangle \cap C_i)}{\mu^*(C_i)} \right) \quad (2.5)$$

In intuitive terms, Eq. 2.5 says that the more examples are given, the greater the likelihood for hypotheses that suggest categories that on average overlap well with the examples. This modification captures the role of exemplar variability – the added value from T&G's expansion of Shepard's original model of categorisation. It not only considers the size of a region, but also takes into account the relative locations of category examples, and the frequency at which they are observed.

The modification can help a reply to the pragmatic challenge. If one category, C_a is used more frequently in the learning history than another category, C_b, the likelihood for the former must, overall, be greater than the likelihood for the latter. If C_a and C_b are mutually exclusive regions in conceptual space, the likelihoods equal the relative frequencies at which points in these regions are observed. In a set of 10 examples, $E = \langle e_1, \ldots, e_{10} \rangle$, 7 examples are called 'dax' and overlap with C_a and 3 examples are called 'fep' and overlap with C_b. Thus, 'dax' = 7/10 and 'fep' = 3/10. Thus, $Pr(E, \text{'dax'} | E \in C_a) = .7 > Pr(E, \text{'fep'} | E \in C_b) = .3$. This means that in terms of the likelihood, a rational categoriser should favour a hypothesis that points towards C_a as the more commonly used concept.

This approach could be made more precise by the outlined combination of the conceptual spaces model and probabilistic inference. Inferences as to which category any unknown object belongs to could be expressed by a probability density function that runs over the conceptual space as estimated based on the density of the meaning examples. The more labelled examples for one category as opposed to another are given, the higher the probability density for the corresponding area in conceptual space. For instance, upon observing $\langle e_1, \text{'dax'} \rangle$, the difference in the relative overlap with C_a and with C_b might not be significant. Given multiple examples, however, this difference should increase and make the corresponding prediction for observations of any next item to be called 'dax' more precise. If the next two examples for 'dax', e_2 and e_3, are also in C_a then e_2 and e_3 present evidence that confirms the hypothesis that 'dax' refers to the region C_a relatively more than the hypothesis that 'dax' refers to the region C_b in conceptual space. Thus, given the evidence, the probability density for an area confined to C_a should be greater than the probability density for C_b. Following the original intuition behind T&G's size principle, the relative overlap between E and C_a becomes larger over time because C_a is used more frequently throughout the agent's learning history. Thus, even if C_a and C_b would have the same size, C_a might be confirmed better by the evidence over time than C_b.

2.6 Conclusion

This chapter has outlined two conflicting views on the role of generalisation probabilities in perceptual categorisation. On the one hand, there is Shepard's (1987) view that the probability of generalisation is a derivative of perceptual similarities. On the other hand, there is Tenenbaum and Griffiths' (2001) view that the probability of generalisation governs perceived similarities. I have argued that the theory of conceptual spaces (Gärdenfors 2000, 2014) can be used as a semantic mediator between these conflicting views on the role of generalisation probabilities in perceptual categorisation. The wider implication of this approach is that the notion of similarity is conducive to the psychological plausibility of probabilistic models of categorisation and should therefore not be considered irrelevant in the explanation of the generalisation gradient.

Taken together, I have presented three main reasons to consider a geometric size principle valuable. First, it provides a semantically interpretable basis for Bayesian inference. Second, it can accommodate the intuitions behind both T&G's and Shepard's approaches, that categorisation on average sharpens with an increase in similarity and an increasing number of examples such that generalisation becomes more restrictive. This is positive, for example, because it can explain undergeneralisation. Third, the geometric approach provides additional constraints, such as convexity, that could possibly be built into a Bayesian model of categorisation. This would be advantageous to explain, for example, overgeneralisation in category learning.

Future directions call for a way to connect the outlined model with more objective principles of rationality in categorisation and linguistic communication. Further elaboration is also needed with respect to conditionalisation of degrees of beliefs about category membership in conceptual spaces.

Acknowledgements Thanks to Peter Gärdenfors, Alistair Isaac and Mark Sprevak for helpful comments on an earlier draft of this paper. Thanks to Peter Brössel for helpful discussions along the way.

References

Bloom, P. (2002). *How children learn the meanings of words*. Cambridge: MIT Press.

Carnap, R. (1980). A basic system of inductive logic, part II. *Studies in Inductive Logic and Probability, 2*, 7–155.

Carnap, R. (1988). *Meaning and necessity: A study in semantics and modal logic*. Chicago: University of Chicago Press.

Carroll, J. D., & Arabie, P. (1980). Multidimensional scaling. *Annual Review of Psychology, 31*(1), 607–649.

Dautriche, I., Chemla, E., & Christophe, A. (2016). Word learning: Homophony and the distribution of learning exemplars. *Language Learning and Development, 12*(3), 231–251.

Decock, L., Douven, I., & Sznajder, M. (2016). A geometric principle of indifference. *Journal of Applied Logic*, *19*, 54–70.
Edelman, S. (1998). Representation is representation of similarities. *Behavioral and Brain Sciences*, *21*(4), 449–467.
Fairchild, M. D. (2013). *Color appearance models*. Chichester: Wiley.
Gärdenfors, P. (2000). *Conceptual spaces: The geometry of thought*. Cambridge: MIT Press.
Gärdenfors, P. (2014). *Conceptual spaces: The geometry of meaning*. Cambridge: MIT Press.
Goodman, N. (1972). *Problems and projects*. Indianapolis: Bobbs-Merrill.
Jäger, G. (2010). Natural color categories are convex sets. In M. Aloni, H. Bastiaanse, T. de Jager, & K. Schulz (Eds.), *Logic, language and meaning: 17th Amsterdam Colloquium*, Amsterdam, The Netherlands, 16–18 Dec 2009, revised selected papers (pp. 11–20). Berlin/Heidelberg: Springer.
Krumhansl, C. L. (1978). Concerning the applicability of geometric models to similarity data: The interrelationship between similarity and spatial density. *Psychological Review, 85*(5), 445–463.
Kruskal, J. B. (1964). Multidimensional scaling by optimizing goodness of fit to a nonmetric hypothesis. *Psychometrika, 29*(1), 1–27.
Newton, I. (1687/1934). Principia mathematica. *Newton's Principia, 634*.
Poth, N. L., & Brössel, P. (2015). *A Bayesian answer to the complex-first paradox* (unpublished paper). Salzburg Conference for Young Analytic Philosophy, University of Salzburg.
Rosch, E., & Mervis, C. B. (1975). Family resemblances: Studies in the internal structure of categories. *Cognitive Psychology, 7*(4), 573–605.
Russell, S. J., & Norvig, P. (2002). *Artificial intelligence: A modern approach* (International Edition). Englewood Cliffs: Prentice Hall Pearson.
Shepard, R. N. (1962). The analysis of proximities: Multidimensional scaling with an unknown distance function. I. *Psychometrika, 27*(2), 125–140.
Shepard, R. N. (1981). Psychophysical complementarity. In M. Kubovy, & J. R. Pomerantz (Eds.), *Perceptual organization*. Hillsdale: Erlbaum.[SK].
Shepard, R. N. (1987). Toward a universal law of generalization for psychological science. *Science, 237*(4820), 1317–1323.
Shepard, R. N., & Chipman, S. (1970). Second-order isomorphism of internal representations: Shapes of states. *Cognitive Psychology, 1*(1) 1–17.
Teller, P. (1973). Conditionalization and observation. *Synthese, 26*(2), 218–258.
Tenenbaum, J. B., & Griffiths, T. L. (2001). Generalization, similarity, and bayesian inference. *Behavioral and Brain Sciences, 24*, 629–640.
Tversky, A. (1977) Features of similarity. *Psychological Review, 84*(4), 327.
Xu, F., & Tenenbaum, J. B. (2007). Word learning as bayesian inference. *Psychological Review, 114*(2), 245.

Chapter 3
Formalized Conceptual Spaces with a Geometric Representation of Correlations

Lucas Bechberger and Kai-Uwe Kühnberger

Abstract The highly influential framework of conceptual spaces provides a geometric way of representing knowledge. Instances are represented by points in a similarity space and concepts are represented by convex regions in this space. After pointing out a problem with the convexity requirement, we propose a formalization of conceptual spaces based on fuzzy star-shaped sets. Our formalization uses a parametric definition of concepts and extends the original framework by adding means to represent correlations between different domains in a geometric way. Moreover, we define various operations for our formalization, both for creating new concepts from old ones and for measuring relations between concepts. We present an illustrative toy-example and sketch a research project on concept formation that is based on both our formalization and its implementation.

3.1 Introduction

One common criticism of symbolic AI approaches is that the symbols they operate on do not contain any meaning: For the system, they are just arbitrary tokens that can be manipulated in some way. This lack of inherent meaning in abstract symbols is called the symbol grounding problem (Harnad 1990). One approach towards solving this problem is to devise a grounding mechanism that connects abstract symbols to the real world, i.e., to perception and action.

The framework of conceptual spaces (Gärdenfors 2000, 2014) attempts to bridge this gap between symbolic and subsymbolic AI by proposing an intermediate conceptual layer based on geometric representations. A conceptual space is a similarity space spanned by a number of quality dimensions that are based on perception and/or subsymbolic processing. Convex regions in this space correspond to concepts. Abstract symbols can thus be grounded in reality by linking them to regions in a conceptual space whose dimensions are based on perception.

L. Bechberger (✉) · K.-U. Kühnberger
Institute of Cognitive Science, Osnabrück University, Osnabrück, Germany
e-mail: lucas.bechberger@uni-osnabrueck.de; kai-uwe.kuehnberger@uni-osnabrueck.de

M. Kaipainen et al. (eds.), *Conceptual Spaces: Elaborations and Applications*, Synthese Library 405, https://doi.org/10.1007/978-3-030-12800-5_3

The framework of conceptual spaces has been highly influential in the last 15 years within cognitive science and cognitive linguistics (Douven et al. 2011; Fiorini et al. 2013; Warglien et al. 2012; Zenker and Gärdenfors 2015). It has also sparked considerable research in various subfields of artificial intelligence, ranging from robotics and computer vision (Chella et al. 2001, 2003, 2005) over the semantic web and ontology integration (Adams and Raubal 2009b; Dietze and Domingue 2008) to plausible reasoning (Derrac and Schockaert 2015; Schockaert and Prade 2011).

One important question is however left unaddressed by these research efforts: How can an (artificial) agent learn about meaningful regions in a conceptual space purely from unlabeled perceptual data? Our overall approach for solving this concept formation problem is to devise an incremental clustering algorithm that groups a stream of unlabeled observations (represented as points in a conceptual space) into meaningful regions. In this paper, we lay the foundation for this approach by developing a thorough formalization of the conceptual spaces framework.[1]

In Sect. 3.2, we point out that Gärdenfors' convexity requirement prevents a geometric representation of correlations. We resolve this problem by using star-shaped instead of convex sets. Our mathematical formalization presented in Sect. 3.3 defines concepts in a parametric way that is easily implementable. We furthermore define various operations in Sect. 3.4, both for creating new concepts from old ones and for measuring relations between concepts. Moreover, in Sect. 3.5 we describe our implementation of the proposed formalization and illustrate it with a simple conceptual space for fruits. In Sect. 3.6, we summarize other existing formalizations of the conceptual spaces framework and compare them to our proposal. Finally, in Sect. 3.7 we give an outlook on future work with respect to concept formation before concluding the paper in Sect. 3.8.

3.2 Conceptual Spaces

This section presents the cognitive framework of conceptual spaces as described in Gärdenfors (2000) and introduces our formalization of dimensions, domains, and distances. Moreover, we argue that concepts should be represented by star-shaped sets instead of convex sets.

[1] This chapter is a revised and extended version of research presented in the following workshop and conference papers: (Bechberger and Kühnberger 2017a,b,c,d).

3.2.1 Definition of Conceptual Spaces

A conceptual space is a similarity space spanned by a set D of so-called "quality dimensions". Each of these dimensions $d \in D$ represents a cognitively meaningful way in which two stimuli can be judged to be similar or different. Examples for quality dimensions include temperature, weight, time, pitch, and hue. We denote the distance between two points x and y with respect to a dimension d as $|x_d - y_d|$.

A domain $\delta \subseteq D$ is a set of dimensions that inherently belong together. Different perceptual modalities (like color, shape, or taste) are represented by different domains. The color domain for instance consists of the three dimensions hue, saturation, and brightness. Gärdenfors argues based on psychological evidence (Attneave 1950; Shepard 1964) that distance within a domain δ should be measured by the weighted Euclidean metric d_E:

$$d_E^\delta(x, y, W_\delta) = \sqrt{\sum_{d\in\delta} w_d \cdot |x_d - y_d|^2}$$

The parameter W_δ contains positive weights w_d for all dimensions $d \in \delta$ representing their relative importance. We assume that $\sum_{d\in\delta} w_d = 1$.

The overall conceptual space CS can be defined as the product space of all dimensions. Again, based on psychological evidence (Attneave 1950; Shepard 1964), Gärdenfors argues that distance within the overall conceptual space should be measured by the weighted Manhattan metric d_M of the intra-domain distances. Let Δ be the set of all domains in CS. We define the overall distance within a conceptual space as follows:

$$d_C^\Delta(x, y, W) := \sum_{\delta\in\Delta} w_\delta \cdot d_E^\delta(x, y, W_\delta) = \sum_{\delta\in\Delta} w_\delta \cdot \sqrt{\sum_{d\in\delta} w_d \cdot |x_d - y_d|^2}$$

The parameter $W = \langle W_\Delta, \{W_\delta\}_{\delta\in\Delta}\rangle$ contains W_Δ, the set of positive domain weights w_δ. We require that $\sum_{\delta\in\Delta} w_\delta = |\Delta|$. Moreover, W contains for each domain $\delta \in \Delta$ a set W_δ of dimension weights as defined above. The weights in W are not globally constant, but depend on the current context. One can easily show that $d_C^\Delta(x, y, W)$ with a given W is a metric.

The similarity of two points in a conceptual space is inversely related to their distance. Gärdenfors expresses this as follows:

$$Sim(x, y) = e^{-c\cdot d(x,y)} \quad \text{with a constant } c > 0 \text{ and a given metric } d$$

Betweenness is a logical predicate $B(x, y, z)$ that is true if and only if y is considered to be between x and z. It can be defined based on a given metric d:

$$B_d(x, y, z) :\Longleftrightarrow d(x, y) + d(y, z) = d(x, z)$$

The betweenness relation based on the Euclidean metric d_E results in the straight line segment connecting the points x and z, whereas the betweenness relation based on the Manhattan metric d_M results in an axis-parallel cuboid between the points x and z. We can define convexity and star-shapedness based on the notion of betweenness:

Definition 1 (Convexity) A set $C \subseteq CS$ is *convex* under a metric $d \ :\Longleftrightarrow$
$$\forall x \in C, z \in C, y \in CS : (B_d(x, y, z) \rightarrow y \in C)$$

Definition 2 (Star-shapedness) A set $S \subseteq CS$ is *star-shaped* under a metric d with respect to a set $P \subseteq S \ :\Longleftrightarrow$
$$\forall p \in P, z \in S, y \in CS : (B_d(p, y, z) \rightarrow y \in S)$$

Convexity under the Euclidean metric d_E is equivalent to the common definition of convexity in Euclidean spaces. The only sets that are considered convex under the Manhattan metric d_M are axis-parallel cuboids.

Gärdenfors distinguishes properties like "red", "round", and "sweet" from full-fleshed concepts like "apple" or "dog" by observing that properties can be defined on individual domains (e.g., color, shape, taste), whereas full-fleshed concepts involve multiple domains.

Definition 3 (Property, Gärdenfors 2000) A *natural property* is a convex region of a domain in a conceptual space.

Full-fleshed concepts can be expressed as a combination of properties from different domains. These domains might have a different importance for the concept which is reflected by so-called "salience weights". Another important aspect of concepts are the correlations between the different domains (Medin and Shoben 1988), which are important for both learning (Billman and Knutson 1996) and reasoning (Murphy 2002, Chapter 8).

Definition 4 (Concept, Gärdenfors 2000) A *natural concept* is represented as a set of convex regions in a number of domains together with an assignment of salience weights to the domains and information about how the regions in different domains are correlated.

3.2.2 An Argument Against Convexity

Gärdenfors (2000) does not propose any concrete way for representing correlations between domains. As the main idea of the conceptual spaces framework is to find a geometric representation of conceptual structures, we think that a geometric representation of these correlations is desirable.

Consider the left part of Fig. 3.1. In this example, we consider two domains, age and height, in order to define the concepts of child and adult. We expect a strong correlation between age and height for children, but no such correlation for adults.

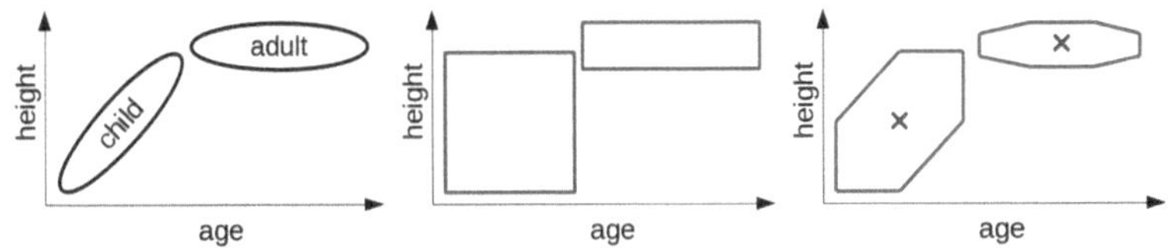

Fig. 3.1 Left: Intuitive way to define regions for the concepts of "adult" and "child". Middle: Representation by using convex sets. Right: Representation by using star-shaped sets with central points marked by crosses

This is represented by the two ellipses.[2] As one can see, the values of age and height constrain each other: For instance, if the value on the age dimension is low and the point lies in the child region, then also the value on the height dimension must be low.

Domains are combined by using the Manhattan metric and convex sets under the Manhattan metric are axis-parallel cuboids. As the dimensions of age and height belong to different domains, a convex representation of the two concepts results in the rectangles shown in the middle of Fig. 3.1. As one can see, all information about the correlation of age and height in the child concept is lost in this representation: The values of age and height do not constrain each other at all. According to this representation, a child of age 2 with a height of 1.60 m would be totally conceivable. This illustrates that we cannot geometrically represent correlations between domains if we assume that concepts are convex and that the Manhattan metric is used. We think that our example is not a pathological one and that similar problems will occur quite frequently when encoding concepts.[3]

Star-shapedness is a weaker constraint than convexity. If we only require concepts to be star-shaped under the Manhattan metric, we can represent the correlation of age and height for children in a geometric way. This is shown in the right part of Fig. 3.1: Both sketched sets are star-shaped under the Manhattan metric with respect to a central point.[4] Although the star-shaped sets do not exactly correspond to our intuitive sketch in the left part of Fig. 3.1, they definitely are an improvement over the convex representation.

By definition, star-shaped sets cannot contain any "holes". They furthermore have a well defined central region P that can be interpreted as a prototype. Thus, the connection that Gärdenfors (2000) established between the prototype

[2]Please note that this is a very simplified artificial example to illustrate our main point.

[3]For instance, there is an obvious correlation between a banana's color and its taste. If you replace the "age" dimension with "hue" and the "height" dimension with "sweetness" in Fig. 3.1, you will observe similar encoding problems for the "banana" concept as for the "child" concept.

[4]Please note that although the sketched sets are still convex under the Euclidean metric, they are star-shaped but not convex under the Manhattan metric.

theory of concepts and the framework of conceptual spaces is preserved. Replacing convexity with star-shapedness is therefore only a minimal departure from the original framework.

The problem illustrated in Fig. 3.1 could also be resolved by replacing the Manhattan metric with a different distance function for combining domains. A natural choice would be to use the Euclidean metric everywhere. We think however that this would be a major modification of the original framework. For instance, if the Euclidean metric is used everywhere, the usage of domains to structure the conceptual space loses its main effect of influencing the overall distance metric. Moreover, there exists some psychological evidence (Attneave 1950; Johannesson 2001; Shepard 1964, 1987) which indicates that human similarity ratings are reflected better by the Manhattan metric than by the Euclidean metric if different domains are involved (e.g., stimuli differing in size and brightness). As a psychologically plausible representation of similarity is one of the core principles of the conceptual spaces framework, these findings should be taken into account. Furthermore, in high-dimensional feature spaces the Manhattan metric provides a better relative contrast between close and distant points than the Euclidean metric (Aggarwal et al. 2001). If we expect a large number of domains, this also supports the usage of the Manhattan metric from an implementational point of view. One could of course also replace the Manhattan metric with something else (e.g., the Mahanalobis distance). However, as there is currently no strong evidence supporting the usage of any particular other metric, we think that it is best to use the metrics proposed in the original conceptual spaces framework.

Based on these arguments, we think that relaxing the convexity constraint is a better option than abolishing the use of the Manhattan metric.

Please note that the example given above is intended to highlight representational problems of the conceptual spaces framework if it is applied to artificial intelligence and if a geometric representation of correlations is desired. We do not make any claims that star-shapedness is a psychologically plausible extension of the original framework and we do not know about any psychological data that could support such a claim.

3.3 A Parametric Definition of Concepts

3.3.1 Preliminaries

Our formalization is based on the following insight:

Lemma 1 *Let $C_1, \dots, C_m$ be convex sets in CS under some metric d and let $P := \bigcap_{i=1}^{m} C_i$. If $P \neq \emptyset$, then $S := \bigcup_{i=1}^{m} C_i$ is star-shaped under d with respect to P.*

Proof Obvious (see also Smith 1968). □

We will use axis-parallel cuboids as building blocks for our star-shaped sets. They are defined in the following way[5]:

Definition 5 (Axis-parallel cuboid) We describe an *axis-parallel cuboid*[6] C as a triple $\langle \Delta_C, p^-, p^+ \rangle$. C is defined on the domains $\Delta_C \subseteq \Delta$, i.e., on the dimensions $D_C := \bigcup_{\delta \in \Delta_C} \delta$. We call p^-, p^+ the support points of C and require that:

$$\forall d \in D_C : p_d^+, p_d^- \notin \{+\infty, -\infty\} \quad \wedge \quad \forall d \in D \setminus D_C : p_d^- := -\infty \wedge p_d^+ := +\infty$$

Then, we define the cuboid C in the following way:

$$C = \left\{x \in CS \mid \forall d \in D : p_d^- \leq x_d \leq p_d^+\right\}$$

Lemma 2 *A cuboid C is convex under d_C^Δ, given a fixed set of weights W.*

Proof It is easy to see that cuboids are convex with respect to d_M and d_E. Based on this, one can show that they are also convex with respect to d_C^Δ, which is a combination of d_M and d_E. □

Our formalization will make use of fuzzy sets (Zadeh 1965), which can be defined in our current context as follows:

Definition 6 (Fuzzy set) A *fuzzy set* $\widetilde{A}$ on CS is defined by its membership function $\mu_{\widetilde{A}} : CS \to [0, 1]$.

For each $x \in CS$, we interpret $\mu_{\widetilde{A}}(x)$ as degree of membership of x in $\widetilde{A}$. Note that fuzzy sets contain crisp sets as a special case where $\mu_{\widetilde{A}} : CS \to \{0, 1\}$.

Definition 7 (Alpha-cut) Given a fuzzy set $\widetilde{A}$ on CS, its *α-cut* $\widetilde{A}^\alpha$ for $\alpha \in [0, 1]$ is defined as follows:

$$\widetilde{A}^\alpha := \left\{x \in CS \mid \mu_{\widetilde{A}}(x) \geq \alpha\right\}$$

Definition 8 (Fuzzy star-shapedness) A fuzzy set $\widetilde{A}$ is called *star-shaped* under a metric d with respect to a crisp set P if and only if all of its α-cuts $\widetilde{A}^\alpha$ are either empty or star-shaped under d with respect to P.

One can also generalize the ideas of subsethood, intersection, and union from crisp to fuzzy sets. We adopt the most widely used definitions:

Definition 9 (Operations on fuzzy sets) Let $\widetilde{A}, \widetilde{B}$ be two fuzzy sets defined on CS.

- Subsethood: $\widetilde{A} \subseteq \widetilde{B} :\iff \left(\forall x \in CS : \mu_{\widetilde{A}}(x) \leq \mu_{\widetilde{B}}(x)\right)$
- Intersection: $\forall x \in CS : \mu_{\widetilde{A} \cap \widetilde{B}}(x) := \min\left(\mu_{\widetilde{A}}(x), \mu_{\widetilde{B}}(x)\right)$
- Union: $\forall x \in CS : \mu_{\widetilde{A} \cup \widetilde{B}}(x) := \max\left(\mu_{\widetilde{A}}(x), \mu_{\widetilde{B}}(x)\right)$

[5] All of the following definitions and propositions hold for any number of dimensions.

[6] We will drop the modifier "axis-parallel" from now on.

3.3.2 Fuzzy Simple Star-Shaped Sets

By combining Lemmas 1 and 2, we see that any union of intersecting cuboids is star-shaped under d_C^{Δ}. We use this insight to define simple star-shaped sets (illustrated in Fig. 3.2 with $m = 3$ cuboids), which will serve as cores for our concepts:

Definition 10 (Simple star-shaped set) We describe a *simple star-shaped set* S as a tuple $\langle \Delta_S, \{C_1, \ldots, C_m\}\rangle$ where $\Delta_S \subseteq \Delta$ is a set of domains on which the cuboids $\{C_1, \ldots, C_m\}$ (and thus also S) are defined. We further require that the central region $P := \bigcap_{i=1}^m C_i \neq \emptyset$. Then the simple star-shaped set S is defined as

$$S := \bigcup_{i=1}^{m} C_i$$

In practice, it is often not possible to define clear-cut boundaries for concepts and properties. It is, for example, very hard to define a generally accepted crisp boundary for the property "red". We therefore use a fuzzified version of simple star-shaped sets for representing concepts, which allows us to define imprecise concept boundaries. This usage of fuzzy sets for representing concepts has already a long tradition (Bělohlávek and Klir 2011; Douven et al. 2011; Osherson and Smith 1982; Ruspini 1991; Zadeh 1982). We use a simple star-shaped set S as a concept's "core" and define the membership of any point $x \in CS$ to this concept as $\max_{y \in S} Sim(x, y)$:

Definition 11 (Fuzzy simple star-shaped set) A *fuzzy simple star-shaped set* $\widetilde{S}$ is described by a quadruple $\langle S, \mu_0, c, W\rangle$ where $S = \langle \Delta_S, \{C_1, \ldots, C_m\}\rangle$ is a non-empty simple star-shaped set. The parameter $\mu_0 \in (0, 1]$ determines the highest possible membership to $\widetilde{S}$ and is usually set to 1. The sensitivity parameter $c > 0$ controls the rate of the exponential decay in the similarity function. Finally, $W = \langle W_{\Delta_S}, \{W_\delta\}_{\delta \in \Delta_S}\rangle$ contains positive weights for all domains in Δ_S and all dimensions within these domains, reflecting their respective importance. We require

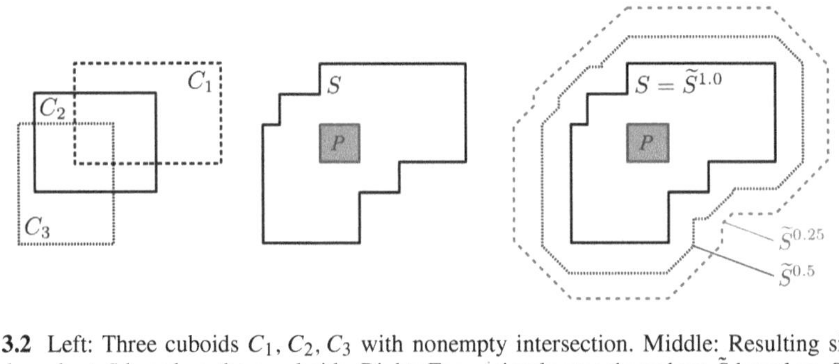

Fig. 3.2 Left: Three cuboids C_1, C_2, C_3 with nonempty intersection. Middle: Resulting simple star-shaped set S based on these cuboids. Right: Fuzzy simple star-shaped set $\widetilde{S}$ based on S with three α-cuts for $\alpha \in \{1.0, 0.5, 0.25\}$

that $\sum_{\delta \in \Delta_S} w_\delta = |\Delta_S|$ and that $\forall \delta \in \Delta_S : \sum_{d \in \delta} w_d = 1$. The membership function of $\widetilde{S}$ is then defined as follows:

$$\mu_{\widetilde{S}}(x) := \mu_0 \cdot \max_{y \in S} \left(e^{-c \cdot d_C^{\Delta_S}(x,y,W)} \right)$$

The sensitivity parameter c controls the overall degree of fuzziness of $\widetilde{S}$ by determining how fast the membership drops to zero (larger values of c result in steeper drops of the membership function). The weights W represent not only the relative importance of the respective domain or dimension for the represented concept, but they also influence the relative fuzziness with respect to this domain or dimension (again, larger weights cause steeper drops). Note that if $|\Delta_S| = 1$, then $\widetilde{S}$ represents a property, and if $|\Delta_S| > 1$, then $\widetilde{S}$ represents a concept.

The right part of Fig. 3.2 shows a fuzzy simple star-shaped set $\widetilde{S}$. In this illustration, the x and y axes are assumed to belong to different domains, and are combined with the Manhattan metric using equal weights.

Lemma 3 *Let $\widetilde{S} = \langle S, \mu_0, c, W \rangle$ be a fuzzy simple star-shaped set and let $\alpha \leq \mu_0$. Then, $\widetilde{S}^\alpha$ is equivalent to an ϵ-neighborhood of S with $\epsilon = -\frac{1}{c} \cdot \ln\left(\frac{\alpha}{\mu_0}\right)$.*

Proof

$$x \in \widetilde{S}^\alpha \iff \mu_{\widetilde{S}}(x) = \mu_0 \cdot \max_{y \in S} \left(e^{-c \cdot d_C^{\Delta_S}(x,y,W)} \right) \geq \alpha \iff e^{-c \cdot \min_{y \in S}\left(d_C^{\Delta_S}(x,y,W)\right)} \geq \frac{\alpha}{\mu_0}$$

$$\iff \min_{y \in S} d_C^{\Delta_S}(x,y,W) \leq -\frac{1}{c} \cdot \ln\left(\frac{\alpha}{\mu_0}\right) =: \epsilon \qquad \square$$

Proposition 1 *Any fuzzy simple star-shaped set $\widetilde{S} = \langle S, \mu_0, c, W \rangle$ is star-shaped with respect to $P = \bigcap_{i=1}^m C_i$ under $d_C^{\Delta_S}$.*

Proof For $\alpha \leq \mu_0$, $\widetilde{S}^\alpha$ is an ϵ-neighborhood of S (Lemma 3). We can define the ϵ-neighborhood of a cuboid C_i under $d_C^{\Delta_S}$ as

$$C_i^\epsilon = \left\{ z \in CS \mid \forall d \in D_S : p_{id}^- - u_d \leq z_d \leq p_{id}^+ + u_d \right\}$$

where the vector u represents the difference between $x \in C_i$ and $z \in C_i^\epsilon$. Thus, u must fulfill the following constraints:

$$\sum_{\delta \in \Delta_S} w_\delta \cdot \sqrt{\sum_{d \in \delta} w_d \cdot (u_d)^2} \leq \epsilon \quad \wedge \quad \forall d \in D_S : u_d \geq 0$$

Let now $x \in C_i, z \in C_i^\epsilon$.

$$\forall d \in D_S : \; x_d = p_{id}^- + a_d \text{ with } a_d \in [0, p_{id}^+ - p_{id}^-]$$
$$z_d = p_{id}^- + b_d \text{ with } b_d \in [-u_d, p_{id}^+ - p_{id}^- + u_d]$$

We know that a point $y \in CS$ is between x and z with respect to $d_C^{\Delta_S}$ if the following condition is true:

$$\begin{aligned} & d_C^{\Delta_S}(x, y, W) + d_C^{\Delta_S}(y, z, W) = d_C^{\Delta_S}(x, z, W) \\ \iff & \forall \delta \in \Delta_S : d_E^{\delta}(x, y, W_\delta) + d_E^{\delta}(y, z, W_\delta) = d_E^{\delta}(x, z, W_\delta) \\ \iff & \forall \delta \in \Delta_S : \exists t \in [0, 1] : \forall d \in \delta : y_d = t \cdot x_d + (1 - t) \cdot z_d \end{aligned}$$

The first equivalence holds because $d_C^{\Delta_S}$ is a weighted sum of Euclidean metrics d_E^{δ}. As the weights are fixed and as the Euclidean metric is obeying the triangle inequality, the equation with respect to $d_C^{\Delta_S}$ can only hold if the equation with respect to d_E^{δ} holds for all $\delta \in \Delta$.

We can thus write the components of y like this:

$$\begin{aligned} \forall d \in D_S : \exists t \in [0, 1] : y_d = t \cdot x_d + (1 - t) \cdot z_d = t \cdot \left(p_{id}^- + a_d\right) + (1 - t)\left(p_{id}^- + b_d\right) \\ = p_{id}^- + t \cdot a_d + (1 - t) \cdot b_d = p_{id}^- + c_d \end{aligned}$$

As $c_d := t \cdot a_d + (1 - t) \cdot b_d \in [-u_d, p_{id}^+ - p_{id}^- + u_d]$, it follows that $y \in C_i^{\epsilon}$. So C_i^{ϵ} is star-shaped with respect to C_i under $d_C^{\Delta_S}$. More specifically, C_i^{ϵ} is also star-shaped with respect to P under $d_C^{\Delta_S}$. Therefore, $S^{\epsilon} = \bigcup_{i=1}^{m} C_i^{\epsilon}$ is star-shaped under $d_C^{\Delta_S}$ with respect to P. Thus, all $\widetilde{S}^{\alpha}$ with $\alpha \leq \mu_0$ are star-shaped under $d_C^{\Delta_S}$ with respect to P. It is obvious that $\widetilde{S}^{\alpha} = \emptyset$ if $\alpha > \mu_0$, so $\widetilde{S}$ is star-shaped according to Definition 8. □

From now on, we will use the terms "core" and "concept" to refer to our definitions of simple star-shaped sets and fuzzy simple star-shaped sets, respectively.

3.4 Operations on Concepts

In this section, we develop a number of operations, which can be used to create new concepts from existing ones and to describe relations between concepts.

3.4.1 Intersection

The intersection of two concepts can be interpreted as logical conjunction: Intersecting "green" with "banana" should result in the concept "green banana".

If we intersect two *cores* S_1 and S_2, we simply need to intersect their cuboids. As an intersection of two cuboids is again a cuboid, the result of intersecting two cores can be described as a union of cuboids. It is simple star-shaped if and only if these resulting cuboids have a nonempty intersection. This is only the case if the central regions P_1 and P_2 of S_1 and S_2 intersect.[7]

However, we would like our intersection to result in a valid core even if $P_1 \cap P_2 = \emptyset$. Thus, when intersecting two cores, we might need to apply some repair mechanism in order to restore star-shapedness.

We propose to extend the cuboids C_i of the intersection in such a way that they meet in some "midpoint" $p^* \in CS$ (e.g., the arithmetic mean of their centers). We create extended versions C_i^* of all C_i by defining their support points like this:

$$\forall d \in D : p_{id}^{-*} := \min\left(p_{id}^{-}, p_d^{*}\right), \quad p_{id}^{+*} := \max\left(p_{id}^{+}, p_d^{*}\right)$$

The intersection of the resulting C_i^* contains at least p^*, so it is not empty. This means that $S' = \left\langle \Delta_{S_1} \cup \Delta_{S_2}, \left\{C_1^*, \ldots, C_m^*\right\}\right\rangle$ is again a simple star-shaped set. We denote this modified intersection (consisting of the actual intersection and the application of the repair mechanism) as $S' = I\left(S_1, S_2\right)$.

We define the intersection of two *concepts* as $\widetilde{S}' = I(\widetilde{S}_1, \widetilde{S}_2) := \left\langle S', \mu_0', c', W' \right\rangle$ with:

- $S' := I\left(\widetilde{S}_1^{\alpha'}, \widetilde{S}_2^{\alpha'}\right)$ (where $\alpha' = \max\left\{\alpha \in [0, 1] : \widetilde{S}_1^{\alpha} \cap \widetilde{S}_2^{\alpha} \neq \emptyset\right\}$)
- $\mu_0' := \alpha'$
- $c' := \min\left(c^{(1)}, c^{(2)}\right)$
- W' with weights defined as follows (where $s, t \in [0, 1]$)[8]:

$$\forall \delta \in \Delta_{S_1} \cap \Delta_{S_2} : \left(\left(w_\delta' := s \cdot w_\delta^{(1)} + (1 - s) \cdot w_\delta^{(2)}\right)\right.$$
$$\left.\wedge \forall d \in \delta : \left(w_d' := t \cdot w_d^{(1)} + (1 - t) \cdot w_d^{(2)}\right)\right)$$
$$\forall \delta \in \Delta_{S_1} \setminus \Delta_{S_2} : \left(\left(w_\delta' := w_\delta^{(1)}\right) \wedge \forall d \in \delta : \left(w_d' := w_d^{(1)}\right)\right)$$
$$\forall \delta \in \Delta_{S_2} \setminus \Delta_{S_1} : \left(\left(w_\delta' := w_\delta^{(2)}\right) \wedge \forall d \in \delta : \left(w_d' := w_d^{(2)}\right)\right)$$

When taking the combination of two somewhat imprecise concepts, the result should not be more precise than any of the original concepts. As the sensitivity parameter c is inversely related to fuzziness, we take the minimum. If a weight is defined for both original concepts, we take a convex combination, and if it is only defined for one of them, we simply copy it. The importance of each dimension and domain to the new concept will thus lie somewhere between its importance with respect to the two original concepts.

---

[7]Note that if the two cores are defined on completely different domains (i.e., $\Delta_{S_1} \cap \Delta_{S_2} = \emptyset$), then their central regions intersect (i.e., $P_1 \cap P_2 \neq \emptyset$), because we can find at least one point in the overall space that belongs to both P_1 and P_2.

[8]In some cases, the normalization constraint of the resulting domain weights might be violated. We can enforce this constraint by manually normalizing the weights afterwards.

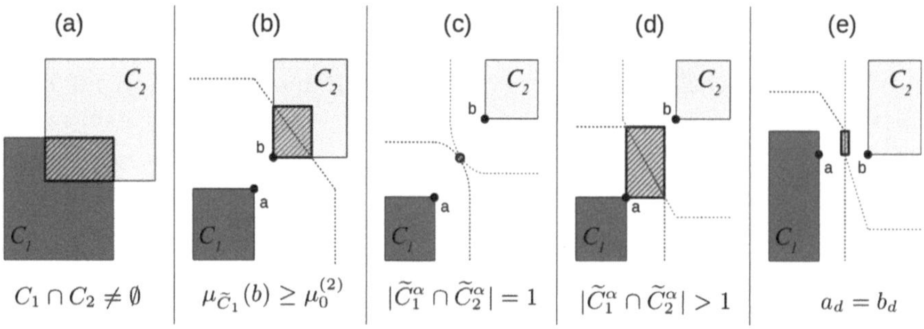

Fig. 3.3 Possible results of intersecting two fuzzy cuboids

The key challenge with respect to the intersection of two concepts $\widetilde{S}_1$ and $\widetilde{S}_2$ is to find the new core S', i.e., the highest non-empty α-cut intersection of the two concepts. We simplify this problem by iterating over all combinations of cuboids $C_1 \in S_1, C_2 \in S_2$ and by looking at each pair of cuboids individually. Let $a \in C_1$ and $b \in C_2$ be the two closest points from the two cuboids under consideration (i.e., $\forall x \in C_1, y \in C_2 : d(a, b) \leq d(x, y)$). Let us define for a cuboid $C \in S$ its fuzzified version $\widetilde{C}$ as follows (cf. Definition 11):

$$\mu_{\widetilde{C}}(x) = \mu_0 \cdot \max_{y \in C} \left(e^{-c \cdot d_C^{\Delta}(x, y, W)} \right)$$

It is obvious that $\mu_{\widetilde{S}}(x) = \max_{C_i \in S} \mu_{\widetilde{C}_i}(x)$. When intersecting two fuzzified cuboids $\widetilde{C}_1$ and $\widetilde{C}_2$, the following results are possible:

1. The crisp cuboids have a nonempty intersection (Fig. 3.3a). In this case, we simply compute their crisp intersection. The α-value of this intersection is equal to $\min(\mu_0^{(1)}, \mu_0^{(2)})$.
2. The μ_0 parameters are different and the $\mu_0^{(i)}$-cut of $\widetilde{C}_j$ intersects with C_i (Fig. 3.3b). In this case, we need to intersect $\widetilde{C}_j^{\mu_0^{(i)}}$ with C_i and approximate the result by a cuboid. The α-value of this intersection is equal to $\mu_0^{(i)}$.
3. The intersection of the two fuzzified cuboids consists of a single point x^* lying between a and b (Fig. 3.3c). In this case, we define a trivial cuboid with $p^- = p^+ = x^*$. The α-value of this intersection is $\mu_{\widetilde{C}_1}(x^*) = \mu_{\widetilde{C}_2}(x^*)$.
4. The intersection of the two fuzzified cuboids consists of a set of points (Fig. 3.3d). This can only happen if the α-cut boundaries of both fuzzified cuboids are parallel to each other, which requires multiple domains to be involved and the weights of both concepts to be linearly dependent. Again, we approximate this intersection by a cuboid. We obtain the α-value of this intersection by computing $\mu_{\widetilde{C}_1}(x) = \mu_{\widetilde{C}_2}(x)$ for some x in the set of points obtained in the beginning.

Moreover, it might happen that both a and b can vary in a certain dimension d (Fig. 3.3e). In this case, the resulting cuboid needs to be extruded in this dimension.

After having computed a cuboid approximation of the intersection for all pairs of fuzzified cuboids, we aggregate them by removing all intersection results with non-maximal α. If the remaining set of cuboids has an empty intersection, we perform the repair mechanism as defined above.

3.4.2 *Unification*

A unification of two or more concepts can be used to construct higher-level concepts. For instance, the concept of "citrus fruit" can be obtained by unifying "lemon", "orange", "grapefruit", "lime", etc.

As each core is defined as a union of cuboids, the unification of two *cores* can also be expressed as a union of cuboids. The resulting set is star-shaped if and only if the central regions of the original cores intersect. So after each unification, we might again need to perform a repair mechanism in order to restore star-shapedness. We propose to use the same repair mechanism that is also used for intersections. We denote the modified unification as $S' = U(S_1, S_2)$.

We define the unification of two *concepts* as $\widetilde{S}' = U(\widetilde{S}_1, \widetilde{S}_2) := \langle S', \mu_0', c', W' \rangle$ with:

- $S' := U(S_1, S_2)$
- $\mu_0' := \max\left(\mu_0^{(1)}, \mu_0^{(2)}\right)$
- c' and W' as described in Sect. 3.4.1

Proposition 2 *Let* $\widetilde{S}_1 = \langle S_1, \mu_0^{(1)}, c^{(1)}, W^{(1)} \rangle$ *and* $\widetilde{S}_2 = \langle S_2, \mu_0^{(2)}, c^{(2)}, W^{(2)} \rangle$ *be two concepts. If we assume that* $\Delta_{S_1} = \Delta_{S_2}$ *and* $W^{(1)} = W^{(2)}$*, then* $\widetilde{S}_1 \cup \widetilde{S}_2 \subseteq U(\widetilde{S}_1, \widetilde{S}_2) = \widetilde{S}'$.

Proof As both $\Delta_{S_1} = \Delta_{S_2}$ and $W^{(1)} = W^{(2)}$, we know that

$$d(x, y) := d_C^{\Delta_{S_1}}\left(x, y, W^{(1)}\right) = d_C^{\Delta_{S_2}}\left(x, y, W^{(2)}\right) = d_C^{\Delta_{S'}}\left(x, y, W'\right)$$

Moreover, $S_1 \cup S_2 \subseteq S'$. Therefore:

$$\begin{aligned}
\mu_{\widetilde{S}_1 \cup \widetilde{S}_2}(x) &= \max\left(\mu_{\widetilde{S}_1}(x), \mu_{\widetilde{S}_2}(x)\right) \\
&= \max\left(\max_{y \in S_1}\left(\mu_0^{(1)} \cdot e^{-c^{(1)} \cdot d(x,y)}\right), \max_{y \in S_2}\left(\mu_0^{(2)} \cdot e^{-c^{(2)} \cdot d(x,y)}\right)\right) \\
&\leq \mu_0' \cdot \max\left(e^{-c^{(1)} \cdot \min_{y \in S_1} d(x,y)}, e^{-c^{(2)} \cdot \min_{y \in S_2} d(x,y)}\right) \\
&\leq \mu_0' \cdot e^{-c' \cdot \min\left(\min_{y \in S_1} d(x,y),\, \min_{y \in S_2} d(x,y)\right)} \\
&\leq \mu_0' \cdot e^{-c' \cdot \min_{y \in S'} d(x,y)} = \mu_{\widetilde{S}'}(x)
\end{aligned}$$

□

3.4.3 Subspace Projection

Projecting a concept onto a subspace corresponds to focusing on certain domains while completely ignoring others. For instance, projecting the concept "apple" onto the color domain results in a property that describes the typical color of apples.

Projecting a cuboid onto a subspace results in a cuboid. As one can easily see, projecting a *core* onto a subspace results in a valid core. We denote the projection of a core S onto domains $\Delta_{S'} \subseteq \Delta_S$ as $S' = P(S, \Delta_{S'})$.

We define the projection of a *concept* $\widetilde{S}$ onto domains $\Delta_{S'} \subseteq \Delta_S$ as $\widetilde{S}' = P(\widetilde{S}, \Delta_{S'}) := \langle S', \mu_0', c', W' \rangle$ with:

- $S' := P(S, \Delta_{S'})$
- $\mu_0' := \mu_0$
- $c' := c$
- $W' := \left\langle \left\{ |\Delta_S'| \cdot \frac{w_\delta}{\sum_{\delta' \in \Delta_{S'}} w_{\delta'}} \right\}_{\delta \in \Delta_{S'}}, \{W_\delta\}_{\delta \in \Delta_{S'}} \right\rangle$

Note that we only apply minimal changes to the parameters: μ_0 and c stay the same, only the domain weights are updated in order to not violate their normalization constraint.

Projecting a set onto two complementary subspaces and then intersecting these projections again in the original space yields a superset of the original set. This is intuitively clear for cores and can also be shown for concepts under one additional constraint:

Proposition 3 *Let $\widetilde{S} = \langle S, \mu_0, c, W \rangle$ be a concept. Let $\widetilde{S}_1 = P(\widetilde{S}, \Delta_1)$ and $\widetilde{S}_2 = P(\widetilde{S}, \Delta_2)$ with $\Delta_1 \cup \Delta_2 = \Delta_S$ and $\Delta_1 \cap \Delta_2 = \emptyset$. Let $\widetilde{S}' = I(\widetilde{S}_1, \widetilde{S}_2)$ as described in Sect.* 3.4.1. *If $\sum_{\delta \in \Delta_1} w_\delta = |\Delta_1|$ and $\sum_{\delta \in \Delta_2} w_\delta = |\Delta_2|$, then $\widetilde{S} \subseteq \widetilde{S}'$.*

Proof We already know that $S \subseteq I(P(S, \Delta_1), P(S, \Delta_2)) = S'$. Moreover, one can easily see that $\mu_0' = \mu_0$ and $c' = c$.

$$\mu_{\widetilde{S}}(x) = \max_{y \in S} \left(\mu_0 \cdot e^{-c \cdot d_C^{\Delta_S}(x,y,W)} \right) \stackrel{!}{\leq} \max_{y \in S'} \left(\mu_0' \cdot e^{-c' \cdot d_C^{\Delta_S}(x,y,W')} \right) = \mu_{\widetilde{S}'}(x)$$

This holds if and only if $W = W'$. $W^{(1)}$ only contains weights for Δ_1, whereas $W^{(2)}$ only contains weights for Δ_2. As $\sum_{\delta \in \Delta_i} w_\delta = |\Delta_i|$ (for $i \in \{1, 2\}$), the weights are not changed during the projection. As $\Delta_1 \cap \Delta_2 = \emptyset$, they are also not changed during the intersection, so $W' = W$. □

3.4.4 Axis-Parallel Cut

In a concept formation process, it might happen that over-generalized concepts are learned (e.g., a single concept that represents both dogs and cats). If it becomes apparent that a finer-grained conceptualization is needed, the system needs to be able to split its current concepts into multiple parts.

One can split a concept $\widetilde{S} = \langle S, \mu_0, c, W \rangle$ into two parts by selecting a value v on a dimension d and by splitting each cuboid $C \in S$ into two child cuboids $C^{(+)} := \{x \in C \mid x_d \geq v\}$ and $C^{(-)} := \{x \in C \mid x_d \leq v\}$.[9] Both $S^{(+)} := \bigcup_{i=1}^{m} C_i^{(+)}$ and $S^{(-)} := \bigcup_{i=1}^{m} C_i^{(-)}$ are still valid cores: They are both a union of cuboids and one can easily show that the intersection of these cuboids is not empty. We define $\widetilde{S}^{(+)} := \langle S^{(+)}, \mu_0, c, W \rangle$ and $\widetilde{S}^{(-)} := \langle S^{(-)}, \mu_0, c, W \rangle$, both of which are by definition valid concepts. Note that by construction, $S^{(-)} \cup S^{(+)} = S$ and $U(\widetilde{S}^{(-)}, \widetilde{S}^{(+)}) = \widetilde{S}$.

3.4.5 *Concept Size*

The size of a concept gives an intuition about its specificity: Large concepts are more general and small concepts are more specific. This is one obvious aspect in which one can compare two concepts to each other.

One can use a measure M to describe the size of a fuzzy set. It can be defined in our context as follows (cf. Bouchon-Meunier et al. 1996):

Definition 12 A measure M on a conceptual space CS is a function $M : \mathcal{F}(CS) \rightarrow \mathbb{R}_0^+$ with $M(\emptyset) = 0$ and $\widetilde{A} \subseteq \widetilde{B} \Rightarrow M(\widetilde{A}) \leq M(\widetilde{B})$, where $\mathcal{F}(CS)$ is the fuzzy power set of CS.

A common measure for fuzzy sets is the integral over the set's membership function, which is equivalent to the Lebesgue integral over the fuzzy set's α-cuts:

$$M\left(\widetilde{A}\right) := \int_{CS} \mu_{\widetilde{A}}(x)\, dx = \int_0^1 V\left(\widetilde{A}^{\alpha}\right) d\alpha \tag{3.1}$$

We use $V(\widetilde{A}^{\alpha})$ to denote the volume of a fuzzy set's α-cut. One can easily see that we can use the inclusion-exclusion formula (cf. e.g., Bogart 1989) to compute the overall measure of a concept $\widetilde{S}$ based on the measure of its fuzzified cuboids[10]:

$$M\left(\widetilde{S}\right) = \sum_{l=1}^{m} \left((-1)^{l+1} \cdot \sum_{\substack{\{i_1,\ldots,i_l\} \\ \subseteq \{1,\ldots,m\}}} M\left(\bigcap_{i \in \{i_1,\ldots,i_l\}} \widetilde{C}_i \right) \right) \tag{3.2}$$

The outer sum iterates over the number of cuboids under consideration (with m being the total number of cuboids in S) and the inner sum iterates over all sets of

[9] A strict inequality in the definition of $C^{(+)}$ or $C^{(-)}$ would not yield a cuboid.

[10] Note that the intersection of two overlapping fuzzified cuboids is again a fuzzified cuboid.

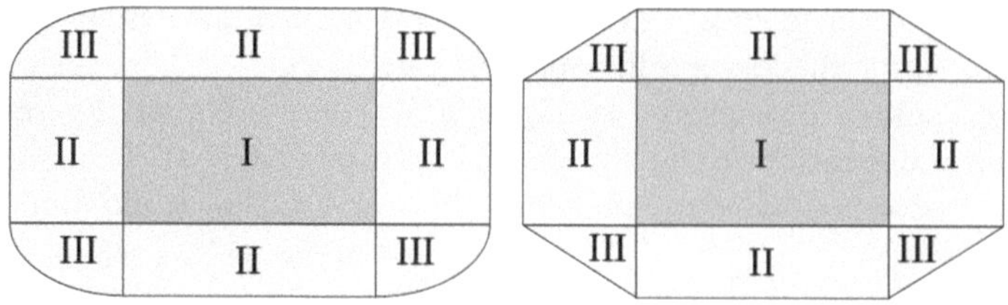

Fig. 3.4 α-cut of a fuzzified cuboid under d_E (left) and d_M (right), respectively

exactly l cuboids. The overall formula generalizes the observation that $|A \cup B| = |A| + |B| - |A \cap B|$ from two to m sets.

In order to derive $M(\widetilde{C})$, we first describe how to compute $V(\widetilde{C}^\alpha)$, i.e., the size of a fuzzified cuboid's α-cut. Using Eq. 3.1, we can then derive $M(\widetilde{C})$, which we can in turn insert into Eq. 3.2 to compute the overall size of $\widetilde{S}$.

Figure 3.4 illustrates the α-cut of a fuzzified two-dimensional cuboid both under d_E (left) and under d_M (right). From Lemma 3 we know that one can interpret each α-cut as an ϵ-neighborhood of the original C with $\epsilon = -\frac{1}{c} \cdot \ln\left(\frac{\alpha}{\mu_0}\right)$.

$V(\widetilde{C}^\alpha)$ can be described as a sum of different components. Let us use the shorthand notation $b_d := p_d^+ - p_d^-$. Looking at Fig. 3.4, one can see that all components of $V(\widetilde{C}^\alpha)$ can be described by ellipses[11]: Component I is a zero-dimensional ellipse (i.e., a point) that was extruded in two dimensions with extrusion lengths of b_1 and b_2, respectively. Component II consists of two one-dimensional ellipses (i.e., line segments) that were extruded in one dimension, and component III is a two-dimensional ellipse.

Let us denote by $\Delta_{\{d_1,\ldots,d_i\}}$ the domain structure obtained by eliminating from Δ all dimensions $d \in D \setminus \{d_1, \ldots, d_i\}$. Moreover, let $V(r, \Delta, W)$ be the hypervolume of a hyperball under $d_C^\Delta(\cdot, \cdot, W)$ with radius r. In this case, a hyperball is the set of all points with a distance of at most r (measured by $d_C^\Delta(\cdot, \cdot, W)$) to a central point. Note that the weights W can cause this ball to have the form of an ellipse. For instance, in Fig. 3.4, we assume that $w_{d_1} < w_{d_2}$ which means that we allow larger differences with respect to d_1 than with respect to d_2. This causes the hyperballs to be stretched in the d_1 dimension, thus obtaining the shape of an ellipse. We can in general describe $V(\widetilde{C}^\alpha)$ as follows:

$$V\left(\widetilde{C}^\alpha\right) = \sum_{i=0}^{n} \left(\sum_{\substack{\{d_1,\ldots,d_i\} \\ \subseteq D}} \left(\prod_{\substack{d \in \\ D \setminus \{d_1,\ldots,d_i\}}} b_d \right) \cdot V\left(-\frac{1}{c} \cdot \ln\left(\frac{\alpha}{\mu_0}\right), \Delta_{\{d_1,\ldots,d_i\}}, W \right) \right)$$

[11] Note that ellipses under d_M have the form of stretched diamonds.

The outer sum of this formula runs over the number of dimensions with respect to which a given point $x \in \widetilde{C}^\alpha$ lies outside of C. We then sum over all combinations $\{d_1, \ldots, d_i\}$ of dimensions for which this could be the case, compute the volume $V(\cdot, \cdot, \cdot)$ of the i-dimensional hyperball in these dimensions, and extrude this intermediate result in all remaining dimensions by multiplying with $\prod_{d \in D \setminus \{d_1,\ldots,d_i\}} b_d$.

Let us illustrate this formula for the α-cuts shown in Fig. 3.4: For $i = 0$, we can only select the empty set for the inner sum, so we end up with $b_1 \cdot b_2$, which is the size of the original cuboid (i.e., component I). For $i = 1$, we can either pick $\{d_1\}$ or $\{d_2\}$ in the inner sum. For $\{d_1\}$, we compute the size of the left and right part of component II by multiplying $V(-\frac{1}{c} \cdot \ln(\frac{\alpha}{\mu_0}), \Delta_{\{d_1\}}, W)$ (i.e., their combined width) with b_2 (i.e., their height). For $\{d_2\}$, we analogously compute the size of the upper and the lower part of component II. Finally, for $i = 2$, we can only pick $\{d_1, d_2\}$ in the inner sum, leaving us with $V(-\frac{1}{c} \cdot \ln(\frac{\alpha}{\mu_0}), \Delta, W)$, which is the size of component III. One can easily see that the formula for $V(\widetilde{C}^\alpha)$ also generalizes to higher dimensions.

As we have shown in Bechberger (2017), $V(r, \Delta, W)$ can be computed as follows, where $n_\delta = |\delta|$:

$$V(r, \Delta, W) = \frac{1}{\prod_{\delta \in \Delta} w_\delta \cdot \prod_{d \in \delta} \sqrt{w_d}} \cdot \frac{r^n}{n!} \cdot \prod_{\delta \in \Delta} \left(n_\delta ! \cdot \frac{\pi^{\frac{n_\delta}{2}}}{\Gamma\left(\frac{n_\delta}{2} + 1\right)} \right)$$

Defining $\delta(d)$ as the unique $\delta \in \Delta$ with $d \in \delta$, and $a_d := w_{\delta(d)} \cdot \sqrt{w_d} \cdot b_d \cdot c$, we can use this observation to rewrite $V(\widetilde{C}^\alpha)$:

$$V\left(\widetilde{C}^\alpha\right) = \frac{1}{c^n \prod_{d \in D} w_{\delta(d)} \sqrt{w_d}} \sum_{i=0}^{n} \left(\frac{(-1)^i \cdot \ln\left(\frac{\alpha}{\mu_0}\right)^i}{i!} \cdot \sum_{\substack{\{d_1,\ldots,d_i\} \\ \subseteq D}} \left(\prod_{\substack{d \in \\ D \setminus \{d_1,\ldots,d_i\}}} a_d \right) \cdot \prod_{\substack{\delta \in \\ \Delta_{\{d_1,\ldots,d_i\}}}} \left(n_\delta ! \cdot \frac{\pi^{\frac{n_\delta}{2}}}{\Gamma\left(\frac{n_\delta}{2} + 1\right)} \right) \right)$$

We can now solve Eq. 3.1 to compute $M(\widetilde{C})$ by using the following lemma:

Lemma 4 $\forall n \in \mathbb{N} : \int_0^1 \ln(x)^n \, dx = (-1)^n \cdot n!$

Proof Substitute $x = e^t$ and $s = -t$, then apply the definition of the Γ function. □

Proposition 4 *The measure of a fuzzified cuboid $\widetilde{C}$ can be computed as follows:*

$$M\left(\widetilde{C}\right) = \frac{\mu_0}{c^n \prod_{d \in D} w_{\delta(d)} \sqrt{w_d}} \sum_{i=0}^{n} \left(\sum_{\substack{\{d_1,\dots,d_i\} \\ \subseteq D}} \left(\prod_{\substack{d \in \\ D \setminus \{d_1,\dots,d_i\}}} a_d \right) \cdot \prod_{\substack{\delta \in \\ \Delta_{\{d_1,\dots,d_i\}}}} \left(n_\delta ! \cdot \frac{\pi^{\frac{n_\delta}{2}}}{\Gamma\left(\frac{n_\delta}{2} + 1\right)} \right) \right)$$

Proof Substitute $x = \frac{\alpha}{\mu_0}$ in Eq. 3.1 and apply Lemma 4. □

Although the formula for $M(\widetilde{C})$ is quite complex, it can be easily implemented via a set of nested loops. As mentioned earlier, we can use the result from Proposition 4 in combination with the inclusion-exclusion formula (Eq. 3.2) to compute $M(\widetilde{S})$ for any concept $\widetilde{S}$. Also Eq. 3.2 can be easily implemented via a set of nested loops. Note that $M(\widetilde{S})$ is always computed only on Δ_S, i.e., the set of domains on which $\widetilde{S}$ is defined.

3.4.6 Subsethood

In order to represent knowledge about a hierarchy of concepts, one needs to be able to determine whether one concept is a subset of another concept. For instance, the fact that $\widetilde{S}_{GrannySmith} \subseteq \widetilde{S}_{apple}$ indicates that Granny Smith is a hyponym of apple.[12]
The classic definition of subsethood for fuzzy sets reads as follows (cf. Definition 9):

$$\widetilde{A} \subseteq \widetilde{B} :\iff \forall x \in CS : \mu_{\widetilde{A}}(x) \leq \mu_{\widetilde{B}}(x)$$

This definition has the weakness of only providing a binary/crisp notion of subsethood. It is desirable to define a *degree* of subsethood in order to make more fine-grained distinctions. Many of the definitions for degrees of subsethood proposed in the fuzzy set literature (Bouchon-Meunier et al. 1996; Young 1996) require that the underlying universe is discrete. The following definition (Kosko 1992) works also in a continuous space and is conceptually quite straightforward:

$$Sub\left(\widetilde{A}, \widetilde{B}\right) = \frac{M\left(\widetilde{A} \cap \widetilde{B}\right)}{M\left(\widetilde{A}\right)} \quad \text{with a measure } M$$

[12] One could also say that the fuzzified cuboids $\widetilde{C}_i$ are sub-concepts of $\widetilde{S}$, because $\widetilde{C}_i \subseteq \widetilde{S}$.

One can interpret this definition intuitively as the "percentage of $\widetilde{A}$ that is also in $\widetilde{B}$". It can be easily implemented based on the intersection defined in Sect. 3.4.1 and the measure defined in Sect. 3.4.5:

$$Sub\left(\widetilde{S}_1, \widetilde{S}_2\right) := \frac{M\left(I\left(\widetilde{S}_1, \widetilde{S}_2\right)\right)}{M\left(\widetilde{S}_1\right)}$$

If $\widetilde{S}_1$ and $\widetilde{S}_2$ are not defined on the same domains, then we first project them onto their shared subset of domains before computing their degree of subsethood.

When computing the intersection of two concepts with different sensitivity parameters $c^{(1)}, c^{(2)}$ and different weights $W^{(1)}, W^{(2)}$, one needs to define new parameters c' and W' for the resulting concept. In Sect. 3.4.1, we have argued that the sensitivity parameter c' should be set to the minimum of $c^{(1)}$ and $c^{(2)}$. Now if $c^{(1)} > c^{(2)}$, then $c' = \min\left(c^{(1)}, c^{(2)}\right) = c^{(2)} < c^{(1)}$. It might thus happen that $M(I(\widetilde{S}_1, \widetilde{S}_2)) > M(\widetilde{S}_1)$, and that therefore $Sub(\widetilde{S}_1, \widetilde{S}_2) > 1$. As we would like to confine $Sub(\widetilde{S}_1, \widetilde{S}_2)$ to the interval $[0, 1]$, we should use the same c and W for computing both $M(I(\widetilde{S}_1, \widetilde{S}_2))$ and $M(\widetilde{S}_1)$.

When judging whether $\widetilde{S}_1$ is a subset of $\widetilde{S}_2$, we can think of $\widetilde{S}_2$ as setting the context by determining the relative importance of the different domains and dimensions as well as the degree of fuzziness. For instance, when judging whether tomatoes are vegetables, we focus our attention on the features that are crucial to the definition of the "vegetable" concept. We thus propose to use $c^{(2)}$ and $W^{(2)}$ when computing $M(I(\widetilde{S}_1, \widetilde{S}_2))$ and $M(\widetilde{S}_1)$.

3.4.7 Implication

Implications play a fundamental role in rule-based systems and all approaches that use formal logics for knowledge representation. It is therefore desirable to define an implication function on concepts, such that one is able to express facts like $apple \Rightarrow red$ within our formalization.

In the fuzzy set literature (Mas et al. 2007), a fuzzy implication is defined as a generalization of the classical crisp implication. Computing the implication of two fuzzy sets typically results in a new fuzzy set which describes the local validity of the implication for each point in the space. In our setting, we are however more interested in a single number that indicates the overall validity of the implication $apple \Rightarrow red$. We propose to reuse the definition of subsethood from Sect. 3.4.6: It makes intuitive sense in our geometric setting to say that $apple \Rightarrow red$ is true to the degree to which *apple* is a subset of *red*. We therefore define:

$$Impl\left(\widetilde{S}_1, \widetilde{S}_2\right) := Sub\left(\widetilde{S}_1, \widetilde{S}_2\right)$$

3.4.8 *Similarity and Betweenness*

Similarity and betweenness of concepts can be valuable sources of information for common-sense reasoning (Derrac and Schockaert 2015): If two concepts are similar, they are expected to have similar properties and behave in similar ways (e.g., pencils and crayons). If one concept (e.g., "master student") is conceptually between two other concepts (e.g., "bachelor student" and "PhD student"), it is expected to share all properties and behaviors that the two other concepts have in common (e.g., having to pay an enrollment fee).

In Sect. 3.2.1, we have already provided definitions for similarity and betweenness of points. We can naively define similarity and betweenness for concepts by applying the definitions from Sect. 3.2.1 to the midpoints of the concepts' central regions P (cf. Definition 10). For computing the similarity, we propose to use both the dimension weights and the sensitivity parameter of the second concept, which again in a sense provides the context for the similarity judgement. If the two concepts are defined on different sets of domains, we use only their common subset of domains for computing the distance of their midpoints and thus their similarity. Betweenness is a binary relation and independent of dimension weights and sensitivity parameters.

These proposed definitions are clearly very naive and shall be replaced by more sophisticated definitions in the future. Especially a graded notion of betweenness would be desirable.

3.5 Implementation and Example

We have implemented our formalization in Python 2.7 and have made its source code publicly available on GitHub[13] (Bechberger 2018). Figure 3.5 shows a class diagram illustrating the overall structure of our implementation. As one can see, each of the components from our definition (i.e., weights, cuboids, cores, and concepts) is represented by an individual class. Moreover, the "cs" module contains the overall domain structure of the conceptual space (represented as a dictionary mapping from domain identifiers to sets of dimensions) along with some utility functions (e.g., computing distance and betweenness of points). The "concept_inspector" package contains a visualization tool that displays 3D and 2D projections of the concepts stored in the "cs" package. When defining a new concept from scratch, one needs to use all of the classes, as all components of the concept need to be specified in detail. When operating with existing concepts, it is however sufficient to use the `Concept` class which contains all the operations defined in Sect. 3.4.

[13]See https://github.com/lbechberger/ConceptualSpaces.

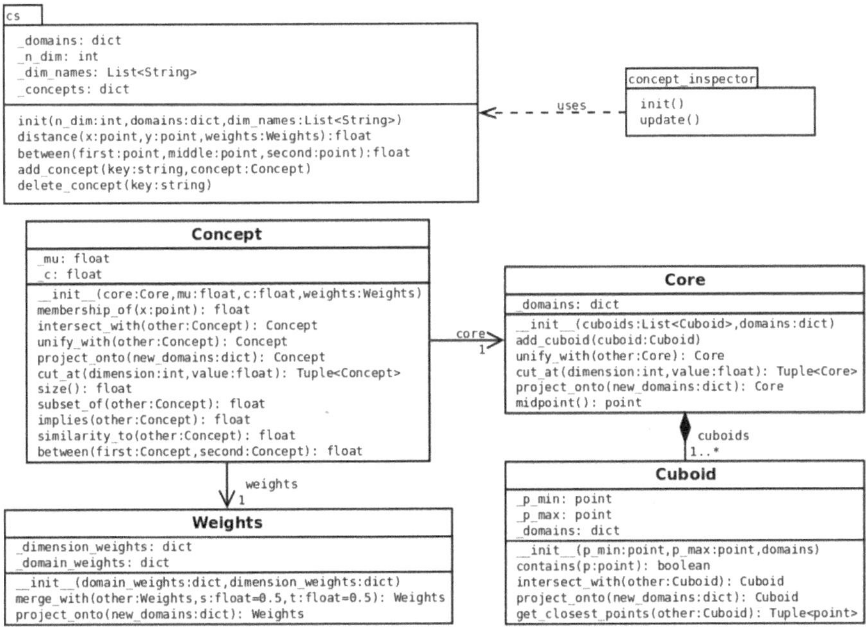

Fig. 3.5 Class diagram of our implementation

Our implementation of the conceptual spaces framework contains a simple toy example – a three-dimensional conceptual space for fruits, defined as follows:

$$\Delta = \left\{\delta_{color} = \{d_{hue}\}, \delta_{shape} = \{d_{round}\}, \delta_{taste} = \{d_{sweet}\}\right\}$$

d_{hue} describes the hue of the observation's color, ranging from 0.00 (purple) to 1.00 (red). d_{round} measures the percentage to which the bounding circle of an object is filled. d_{sweet} represents the relative amount of sugar contained in the fruit, ranging from 0.00 (no sugar) to 1.00 (high sugar content). As all domains are one-dimensional, the dimension weights w_d are always equal to 1.00 for all concepts. We assume that the dimensions are ordered like this: $d_{hue}, d_{round}, d_{sweet}$. The conceptual space is defined as follows in the code:

```
domains = {'color':[0], 'shape':[1], 'taste':[2]}
dimension_names = ['hue', 'round', 'sweet']
space.init(3, domains, dimension_names)
```

Table 3.1 defines some concepts in this space and Fig. 3.6 visualizes them. In the code, concepts can be defined as follows:

```
c_pear = Cuboid([0.5, 0.4, 0.35], [0.7, 0.6, 0.45], domains)
s_pear = Core([c_pear], domains)
w_pear = Weights({'color':0.50, 'shape':1.25, 'taste':1.25},
         {'color':{0:1.0}, 'shape':{1:1.0}, 'taste':{2:1.0}})
pear = Concept(s_pear, 1.0, 12.0, w_pear)
```

Table 3.1 Definitions of several concepts

Concept	Δ_S	p^-	p^+	μ_0	c	W		
						$w_{\delta_{color}}$	$w_{\delta_{shape}}$	$w_{\delta_{taste}}$
Pear	Δ	$(0.50, 0.40, 0.35)$	$(0.70, 0.60, 0.45)$	1.0	12.0	0.50	1.25	1.25
Orange	Δ	$(0.80, 0.90, 0.60)$	$(0.90, 1.00, 0.70)$	1.0	15.0	1.00	1.00	1.00
Lemon	Δ	$(0.70, 0.45, 0.00)$	$(0.80, 0.55, 0.10)$	1.0	20.0	0.50	0.50	2.00
Granny Smith	Δ	$(0.55, 0.70, 0.35)$	$(0.60, 0.80, 0.45)$	1.0	25.0	1.00	1.00	1.00
Apple	Δ	$(0.50, 0.65, 0.35)$ $(0.65, 0.65, 0.40)$ $(0.70, 0.65, 0.45)$	$(0.80, 0.80, 0.50)$ $(0.85, 0.80, 0.55)$ $(1.00, 0.80, 0.60)$	1.0	10.0	0.50	1.50	1.00
Banana	Δ	$(0.50, 0.10, 0.35)$ $(0.70, 0.10, 0.50)$ $(0.75, 0.10, 0.50)$	$(0.75, 0.30, 0.55)$ $(0.80, 0.30, 0.70)$ $(0.85, 0.30, 1.00)$	1.0	10.0	0.75	1.50	0.75
Red	$\{\delta_{color}\}$	$(0.90, -\infty, -\infty)$	$(1.00, +\infty, +\infty)$	1.0	20.0	1.00	–	–

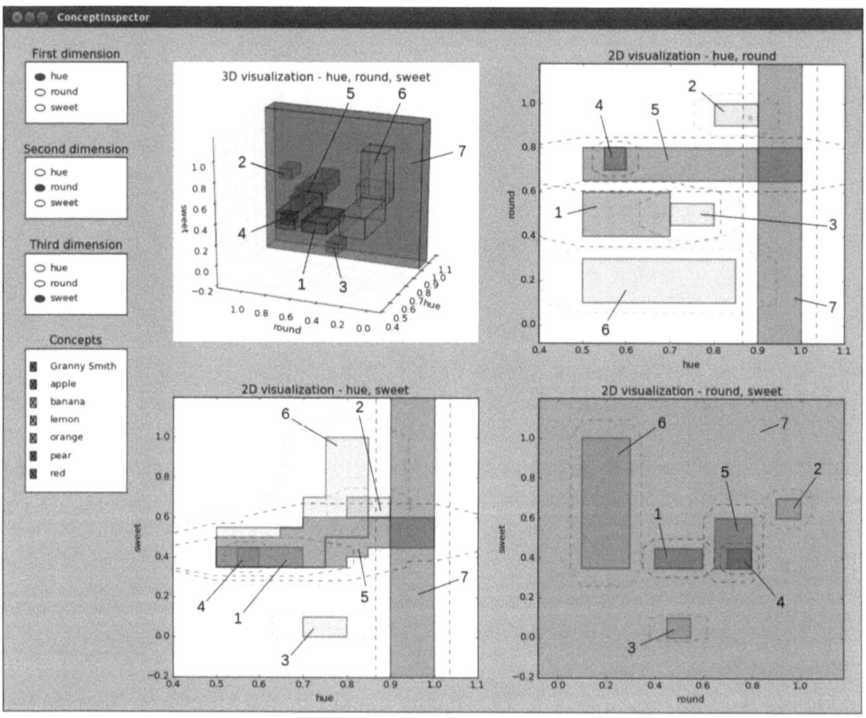

Fig. 3.6 Screenshot of the ConceptInspector tool for the fruit space example with subsequently added labels: pear (1), orange (2), lemon (3), Granny Smith (4), apple (5), banana (6), red (7). The 3D visualization only shows the concepts' cores, the 2D visualizations also illustrate the concepts' 0.5-cuts

We can load the definition of this fruit space into our python interpreter and apply the different operations described in Sect. 3.4 to these concepts. This looks for example as follows:

```
>>> execfile('fruit_space.py')
>>> granny_smith.subset_of(apple)
1.0
>>> apple.implies(red)
0.3333333333333332
>>> (pear.similarity_to(apple), pear.similarity_to(lemon))
(0.007635094218859955, 1.8553913626159717e-07)
>>> print apple.intersect_with(pear)
core: {[0.5, 0.625, 0.35]-[0.7, 0.625, 0.45]}
mu: 0.6872892788
c: 10.0
weights: <{'color': 0.5, 'taste': 1.125, 'shape': 1.375},
    {'color': {0: 1.0}, 'taste': {2: 1.0}, 'shape': {1: 1.0}}>
```

3.6 Related Work

Our work is of course not the first attempt to devise an implementable formalization of the conceptual spaces framework. In this section, we review its strongest competitors.

An early and very thorough formalization was done by Aisbett and Gibbon (2001). Like we, they consider concepts to be regions in the overall conceptual space. However, they stick with Gärdenfors' assumption of convexity and do not define concepts in a parametric way. The only operations they provide are distance and similarity of points and regions. Their formalization targets the interplay of symbols and geometric representations, but it is too abstract to be implementable.

Rickard (2006) and Rickard et al. (2007) provide a formalization based on fuzziness. Their starting points are properties defined in individual domains. They represent concepts as co-occurrence matrices of properties. For instance, $C(sour, green) = 0.8$ in the "apple" concept represents that in 80% of the cases where the property "sour" was present, also the property "green" was observed. By using some mathematical transformations, Rickard et al. interpret these matrices as fuzzy sets on the universe of ordered property pairs. Operations defined on these concepts include similarity judgements between concepts and between concepts and instances. The representation of Rickard et al. nicely captures the correlations between different properties, but their representation of correlations is not geometrical: They first discretize the domains by defining properties and then compute co-occurrence statistics between these properties. Depending on the discretization, this might lead to a relatively coarse-grained notion of correlation. Moreover, as properties and concepts are represented in different ways, one has to use different learning and reasoning mechanisms for them. The formalization by Rickard et al. is also not easy to work with due to the complex mathematical transformations involved.

We would also like to point out that something very similar to the co-occurrence values used by Rickard et al. can be extracted from our representation. One can interpret for instance $C(sour, green)$ as the degree of truth of the implication $\widetilde{S}_{sour} \Rightarrow \widetilde{S}_{green}$ within the apple concept. This number can be computed by using our implication operation:

$$C(sour, green) = Impl\left(I\left(\widetilde{S}_{sour}, \widetilde{S}_{apple}\right), I\left(\widetilde{S}_{green}, \widetilde{S}_{apple}\right)\right)$$

Adams and Raubal (2009a) represent concepts by one convex polytope per domain. This allows for efficient computations while supporting a more fine-grained representation than our cuboid-based approach. The Manhattan metric is used to combine different domains. However, correlations between different domains are not taken into account as each convex polytope is only defined on a single domain. Adams and Raubal also define operations on concepts, namely intersection, similarity computation, and concept combination. This makes their formalization quite similar in spirit to ours.

One could generalize their approach by using polytopes that are defined on the overall space and that are convex under the Euclidean and star-shaped under the Manhattan metric. However, we have found that this requires additional constraints in order to ensure star-shapedness. The number of these constraints grows exponentially with the number of dimensions. Each modification of a concept's description would then involve a large constraint satisfaction problem, rendering this representation unsuitable for learning processes. Our cuboid-based approach is more coarse-grained, but it only involves a single constraint, namely that the intersection of the cuboids is not empty.

Lewis and Lawry (2016) formalize conceptual spaces using random set theory. A random set can be characterized by a set of prototypical points P and a threshold ϵ. Instances that have a distance of at most ϵ to the prototypical set are considered to be elements of the set. The threshold is however not exactly determined, only its probability distribution δ is known. Based on this uncertainty, a membership function $\mu(x)$ can be defined that corresponds to the probability of the distance of a given point x to any prototype $p \in P$ being smaller than ϵ. Lewis and Lawry define properties as random sets within single domains and concepts as random sets in a boolean space whose dimensions indicate the presence or absence of properties. In order to define this boolean space, a single property is taken from each domain. This is in some respect similar to the approach of Rickard (2006) and Rickard et al. (2007) where concepts are also defined on top of existing properties. However, whereas Rickard et al. use two separate formalisms for properties and concepts, Lewis and Lawry use random sets for both (only the underlying space differs).

Lewis and Lawry illustrate how their mathematical formalization is capable of reproducing some effects from the psychological concept combination literature. However, they do not develop a way of representing correlations between domains (such as "red apples are sweet and green apples are sour"). One possible way to do this within their framework would be to define two separate concepts "red apple" and "green apple" and then define on top of them a disjunctive concept "apple = red apple or green apple". This however is a quite indirect way of defining correlations, whereas our approach is intuitively much easier to grasp. Nevertheless, their approach is similar to ours in using a distance-based membership function to a set of prototypical points while using the same representational mechanisms for both properties and concepts.

None of the approaches listed above provides a set of operations that is as comprehensive as the one offered by our proposed formalization. Many practical applications of conceptual spaces (e.g., Chella et al. 2003; Derrac and Schockaert 2015; Dietze and Domingue 2008; Raubal 2004) use only partial ad-hoc implementations of the conceptual spaces framework which usually ignore some important aspects of the framework (e.g., the domain structure).

The only publicly available implementation of the conceptual spaces framework that we are currently aware of is provided by Lieto et al. (2015, 2017). They propose a hybrid architecture that represents concepts by using both description logics and conceptual spaces. This way, symbolic ontological information and similarity-based "common sense" knowledge can be used in an integrated way. Each concept is represented by a single prototypical point and a number of exemplar points. Correlations between domains can therefore only be encoded through the selection of appropriate exemplars. Their work focuses on classification tasks and does therefore not provide any operations for combining different concepts. With respect to the larger number of supported operations, our formalization and implementation can thus be considered more general than theirs. In contrast to our work, the current implementation of their system.[14] comes without any publicly available source code.[15]

3.7 Outlook and Future Work

As stated earlier, our overall research goal is to devise a symbol grounding mechanism by defining a concept formation process in the conceptual spaces framework. Concept formation (Gennari et al. 1989) is the process of incrementally creating a meaningful hierarchical categorization of unlabeled observations. One can easily see that a successful concept formation process implicitly solves the symbol grounding problem: If we are able to find a bottom-up process that can group observations into meaningful categories, these categories can be linked to abstract symbols. These symbols are then grounded in reality, as the concepts they refer to are generalizations of actual observations.

We aim for a three-layered architecture as depicted in Fig. 3.7 with the formalization presented in this paper serving as middle layer. The conceptual space

[14]See http://www.dualpeccs.di.unito.it/download.html.

[15]The source code of an earlier and more limited version of their system can be found here: http://www.di.unito.it/~lieto/cc_classifier.html.

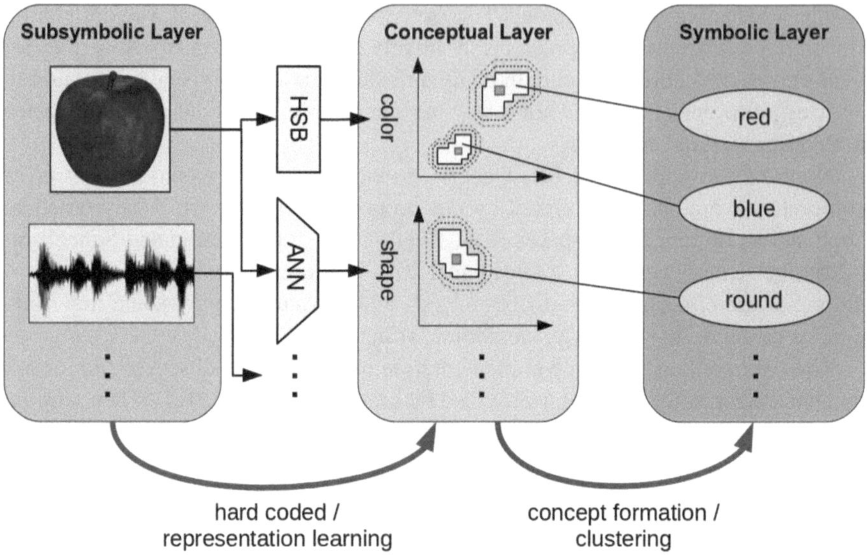

Fig. 3.7 Visualization of our envisioned symbol grounding architecture. The translation from the subsymbolic to the conceptual layer will be based on both hard-coded conversions (like the HSB color space) and pre-trained artificial neural networks. The translation from the conceptual layer to the symbolic layer will be based on a clustering algorithm that performs concept formation

that we will use for our work on concept formation will have a predefined structure: Domains having a well-known structure will be hand-crafted, e.g., color or sound. These domains can quite straightforwardly be represented by a handful of dimensions. Domains with an unclear internal structure (i.e., where it is inherently hard to hand-craft a dimensional representation) can potentially be obtained by deep representation learning (Bengio et al. 2013). This includes for instance the domain of shapes. We will investigate different neural network architectures that have shown promising results in extracting meaningful dimensions from unlabeled data sets, e.g., InfoGAN (Chen et al. 2016) and beta-VAE (Higgins et al. 2017).

This set-up of the conceptual space is not seen as an active part of the concept formation process, but as a preprocessing step to lift information from the subsymbolic layer (e.g., raw pixel information from images) to the conceptual layer.

The most straightforward way of implementing concept formation within the conceptual spaces framework is to use a clustering algorithm that groups unlabeled instances into meaningful regions. Each of these regions can be interpreted as a concept and a symbol can be attached to it.

As this research is seen in the context of artificial general intelligence, it will follow the assumption of insufficient knowledge and resources (Wang 2011): The system will only have *limited resources*, i.e., it will not be able to store all observed data points. This calls for an incremental clustering process. The system will also have to cope with *incomplete information*, i.e., with incomplete feature vectors. As concept hierarchies are an important and useful aspect of human conceptualizations (Murphy 2002, Chapter 7), the concept formation process should ideally result in a hierarchy of concepts. It is generally unknown in the beginning how many concepts will be discovered. Therefore, the number of concepts must be adapted over time.

There are some existing clustering algorithms that partially fulfill these requirements (e.g., CLASSIT, Gennari et al. 1989, or SUSTAIN, Love et al. 2004), but none of them is a perfect fit. We will take inspiration from these algorithms in order to devise a new algorithm that fulfills all the requirements stated above. One can easily see that our formalization is able to support such clustering processes: Concepts can be created and deleted. Modifying the support points of the cuboids in a concept's core results in changes to the concept's position, size, and form. One must however ensure that such modifications preserve the non-emptiness of the cuboids' intersection. Moreover, a concept's form can be changed by modifying the parameters c and W: By changing c, one can control the overall degree of fuzziness, and by changing W, one can control how this fuzziness is distributed among the different domains and dimensions. Two neighboring concepts $\widetilde{S}_1, \widetilde{S}_2$ can be merged into a single cluster by unifying them. A single concept can be split up into two parts by using the axis-parallel cut operation.

Finally, our formalization also supports reasoning processes: (Gärdenfors 2000) argues that adjective-noun combinations like "green apple" or "purple banana" can be expressed by combining properties with concepts. This is supported by our operations of intersection and subspace projection:

In combinations like "green apple", property and concept are compatible. We expect that their cores intersect and that the μ_0 parameter of their intersection is therefore relatively large. In this case, "green" should narrow down the color information associated with the "apple" concept. This can be achieved by simply computing their intersection.

In combinations like "purple banana", property and concept are incompatible. We expect that their cores do not intersect and that the μ_0 parameter of their intersection is relatively small. In this case, "purple" should replace the color information associated with the "banana" concept. This can be achieved by first removing the color domain from the "banana" concept (through a subspace projection) and by then intersecting this intermediate result with "purple".

3.8 Conclusion

In this paper, we proposed a new formalization of the conceptual spaces framework. We aimed to geometrically represent correlations between domains, which led us to consider the more general notion of star-shapedness instead of Gärdenfors' favored constraint of convexity. We defined concepts as fuzzy sets based on intersecting cuboids and a similarity-based membership function. Moreover, we provided a comprehensive set of operations, both for creating new concepts based on existing ones and for measuring relations between concepts. This rich set of operations makes our formalization (to the best of our knowledge) the most thorough and comprehensive formalization of conceptual spaces developed so far.

Our implementation of this formalization and its source code are publicly available and can be used by any researcher interested in conceptual spaces. We think that our implementation can be a good foundation for practical research on conceptual spaces and that it will considerably facilitate research in this area.

In future work, we will provide more thorough definitions of similarity and betweenness for concepts, given that our current definitions are rather naive. A potential starting point for this can be the betwenness relations defined by Derrac and Schockaert (2015). Moreover, we will use the formalization proposed in this paper as a starting point for our research on concept formation as outlined in Sect. 3.7.

References

Adams, B., & Raubal, M. (2009a). A metric conceptual space algebra. In *9th International Conference on Spatial Information Theory* (pp. 51–68). Berlin/Heidelberg: Springer.

Adams, B., & Raubal, M. (2009b). Conceptual space markup language (CSML): Towards the cognitive semantic web. In *2009 IEEE International Conference on Semantic Computing*.

Aggarwal, C. C., Hinneburg, A., & Keim, D. A. (2001). On the surprising behavior of distance metrics in high dimensional space. In *8th International Conference on Database Theory* (pp. 420–434). Berlin/Heidelberg: Springer.

Aisbett, J., & Gibbon, G. (2001). A general formulation of conceptual spaces as a meso level representation. *Artificial Intelligence, 133*(1–2), 189–232.

Attneave, F. (1950). Dimensions of similarity. *The American Journal of Psychology, 63*(4), 516–556.

Bechberger, L. (2017). The size of a hyperball in a conceptual space. https://arxiv.org/abs/1708.05263.

Bechberger, L. (2018). lbechberger/ConceptualSpaces: Version 1.1.0. https://doi.org/10.5281/zenodo.1143978.

Bechberger, L., & Kühnberger, K.-U. (2017a). A comprehensive implementation of conceptual spaces. In *5th International Workshop on Artificial Intelligence and Cognition*.

Bechberger, L., & Kühnberger, K.-U. (2017b). A thorough formalization of conceptual spaces. In G. Kern-Isberner, J. Fürnkranz, & M. Thimm (Eds.), *KI 2017: Advances in Artificial Intelligence: 40th Annual German Conference on AI, Proceedings*, Dortmund, 25–29 Sept 2017 (pp. 58–71). Springer International Publishing.

Bechberger, L., & Kühnberger, K.-U. (2017c). Measuring relations between concepts in conceptual spaces. In M. Bramer & M. Petridis (Eds.), *Artificial intelligence XXXIV: 37th SGAI International Conference on Artificial Intelligence, AI 2017, Proceedings*, Cambridge, 12–14 Dec 2017 (pp. 87–100). Springer International Publishing.

Bechberger, L., & Kühnberger, K.-U. (2017d). Towards grounding conceptual spaces in neural representations. In *12th International Workshop on Neural-Symbolic Learning and Reasoning*.

Bělohlávek, R., & Klir, G. J. (2011). *Concepts and fuzzy logic*. Cambridge: MIT Press.

Bengio, Y., Courville, A., & Vincent, P. (2013). Representation learning: A review and new perspectives. *IEEE Transactions on Pattern Analysis and Machine Intelligence, 35*(8), 1798–1828.

Billman, D., & Knutson, J. (1996). Unsupervised concept learning and value systematicitiy: A complex whole aids learning the parts. *Journal of Experimental Psychology: Learning, Memory, and Cognition, 22*(2), 458–475.

Bogart, K. P. (1989). *Introductory combinatorics* (2nd ed.). Philadelphia: Saunders College Publishing.
Bouchon-Meunier, B., Rifqi, M., & Bothorel, S. (1996). Towards general measures of comparison of objects. *Fuzzy Sets and Systems, 84*(2), 143–153.
Chella, A., Frixione, M., & Gaglio, S. (2001). Conceptual spaces for computer vision representations. *Artificial Intelligence Review, 16*(2), 137–152.
Chella, A., Frixione, M., & Gaglio, S. (2003). Anchoring symbols to conceptual spaces: The case of dynamic scenarios. *Robotics and Autonomous Systems, 43*(2–3), 175–188.
Chella, A., Dindo, H., & Infantino, I. (2005). Anchoring by imitation learning in conceptual spaces. In S. Bandini, & S. Manzoni (Eds.), *AI*IA 2005: Advances in Artificial Intelligence* (pp. 495–506). Berlin/New York: Springer.
Chen, X., Duan, Y., Houthooft, R., Schulman, J., Sutskever, I., & Abbeel, P. (2016). InfoGAN: Interpretable representation learning by information maximizing generative adversarial nets. In D. D. Lee, M. Sugiyama, U. V. Luxburg, I. Guyon, & R. Garnett (Eds.), *Advances in neural information processing systems* (Vol. 29, pp. 2172–2180). Curran Associates, Inc. http://papers.nips.cc/paper/6399-infogan-interpretable-representation-learning-by-information-maximizing-generative-adversarial-nets.pdf
Derrac, J., & Schockaert, S. (2015). Inducing semantic relations from conceptual spaces: A data-driven approach to plausible reasoning. *Artificial Intelligence, 228*, 66–94.
Dietze, S., & Domingue, J. (2008). Exploiting conceptual spaces for ontology integration. In *Data Integration Through Semantic Technology (DIST2008) Workshop at 3rd Asian Semantic Web Conference (ASWC 2008)*.
Douven, I., Decock, L., Dietz, R., & Égré, P. (2011). Vagueness: A conceptual spaces approach. *Journal of Philosophical Logic, 42*(1), 137–160.
Fiorini, S. R., Gärdenfors, P., & Abel, M. (2013). Representing part-whole relations in conceptual spaces. *Cognitive Processing, 15*(2), 127–142.
Gärdenfors, P. (2000). *Conceptual spaces: The geometry of thought*. Cambridge: MIT press.
Gärdenfors, P. (2014). *The geometry of meaning: Semantics based on conceptual spaces*. Cambridge: MIT Press.
Gennari, J. H., Langley, P., & Fisher, D. (1989). Models of incremental concept formation. *Artificial Intelligence, 40*(1–3), 11–61.
Harnad, S. (1990). The symbol grounding problem. *Physica D: Nonlinear Phenomena, 42*(1–3), 335–346.
Higgins, I., Matthey, L., Pal, A., Burgess, C., Glorot, X., Botvinick, M., Mohamed, S., & Lerchner, A. (2017). β-VAE: Learning basic visual concepts with a constrained variational framework. In *5th International Conference on Learning Representations*.
Johannesson, M. (2001). The problem of combining integral and separable dimensions. Technical Report HS-IDA-TR-01-002, University of Skövde, School of Humanities and Informatics.
Kosko, B. (1992). *Neural networks and fuzzy systems: A dynamical systems approach to machine intelligence*. Englewood Cliffs: Prentice Hall.
Lewis, M., & Lawry, J. (2016). Hierarchical conceptual spaces for concept combination. *Artificial Intelligence, 237*, 204–227.
Lieto, A., Minieri, A., Piana, A., & Radicioni, D. P. (2015). A knowledge-based system for prototypical reasoning. *Connection Science, 27*(2), 137–152.
Lieto, A., Radicioni, D. P., & Rho, V. (2017). Dual PECCS: A cognitive system for conceptual representation and categorization. *Journal of Experimental & Theoretical Artificial Intelligence, 29*(2), 433–452.
Love, B. C., Medin, D. L., & Gureckis, T. M. (2004). SUSTAIN: A network model of category learning. *Psychological Review, 111*(2), 309–332.
Mas, M., Monserrat, M., Torrens, J., & Trillas, E. (2007). A survey on fuzzy implication functions. *IEEE Transactions on Fuzzy Systems, 15*(6):1107–1121.
Medin, D. L., & Shoben, E. J. (1988). Context and structure in conceptual combination. *Cognitive Psychology, 20*(2):158–190.
Murphy, G. (2002). *The big book of concepts*. Cambridge: MIT Press.

Osherson, D. N., & Smith, E. E. (1982). Gradedness and conceptual combination. *Cognition, 12*(3), 299–318.
Raubal, M. (2004). Formalizing conceptual spaces. In *Third International Conference on Formal Ontology in Information Systems* (pp. 153–164).
Rickard, J. T. (2006). A concept geometry for conceptual spaces. *Fuzzy Optimization and Decision Making, 5*(4), 311–329.
Rickard, J. T., Aisbett, J., & Gibbon, G. (2007). Knowledge representation and reasoning in conceptual spaces. In *2007 IEEE Symposium on Foundations of Computational Intelligence*.
Ruspini, E. H. (1991). On the semantics of fuzzy logic. *International Journal of Approximate Reasoning, 5*(1), 45–88.
Schockaert, S., & Prade, H. (2011). Interpolation and extrapolation in conceptual spaces: A case study in the music domain. In *5th International Conference on Web Reasoning and Rule Systems* (pp. 217–231). Springer Nature.
Shepard, R. N. (1964). Attention and the metric structure of the stimulus space. *Journal of Mathematical Psychology, 1*(1), 54–87.
Shepard, R. N. (1987). Toward a universal law of generalization for psychological science. *Science, 237*(4820), 1317–1323.
Smith, C. R. (1968). A characterization of star-shaped sets. *The American Mathematical Monthly, 75*(4), 386.
Wang, P. (2011). The assumptions on knowledge and resources in models of rationality. *International Journal of Machine Consciousness, 3*(1), 193–218.
Warglien, M., Gärdenfors, P., & Westera, M. (2012). Event structure, conceptual spaces and the semantics of verbs. *Theoretical Linguistics, 38*(3–4), 159–193.
Young, V. R. (1996). Fuzzy subsethood. *Fuzzy Sets and Systems, 77*(3), 371–384.
Zadeh, L. A. (1965). Fuzzy sets. *Information and Control, 8*(3), 338–353.
Zadeh, L. A. (1982). A note on prototype theory and fuzzy sets. *Cognition, 12*(3), 291–297.
Zenker, F., & Gärdenfors, P. (Eds.). (2015). *Applications of conceptual spaces*. Cham: Springer Science + Business Media.

Chapter 4
Three Levels of Naturalistic Knowledge

Andreas Stephens

Abstract A recent naturalistic epistemological account suggests that there are three nested basic forms of knowledge: procedural knowledge-how, conceptual knowledge-what, and propositional knowledge-that. These three knowledge-forms are grounded in cognitive neuroscience and are mapped to procedural, semantic, and episodic long-term memory respectively. This article investigates and integrates the neuroscientifically grounded account with knowledge-accounts from cognitive ethology and cognitive psychology. It is found that procedural and semantic memory, on a neuroscientific level of analysis, matches an ethological reliabilist account. This formation also matches System 1 from dual process theory on a psychological level, whereas the addition of episodic memory, on the neuroscientific level of analysis, can account for System 2 on the psychological level. It is furthermore argued that semantic memory (conceptual knowledge-what) and the cognitive ability of categorization are linked to each other, and that they can be fruitfully modeled within a conceptual spaces framework.

Keywords Naturalistic epistemology · Cognitive philosophy · Conceptual knowledge · Knowledge-what · Categorization · Conceptual spaces

4.1 Introduction

Investigations regarding knowledge have been going on for millennia while the concept still lacks a sharp and widely accepted definition (see, e.g., Markie 2013; Samet and Zaitchik 2014). However, many philosophers nowadays heed naturalism and consider it the job of science to provide our best explanations. Furthermore, as cognitive sciences have progressed, much relevant information regarding our cognitive faculties and knowledge is indeed available. We understand the world through multiple models, but since different sciences explore cognition

---

A. Stephens (✉)
Lund University, Lund, Sweden

M. Kaipainen et al. (eds.), *Conceptual Spaces: Elaborations and Applications*, Synthese Library 405, https://doi.org/10.1007/978-3-030-12800-5_4

and knowledge on different levels of analysis, it is not clear if, or how, the different accounts of knowledge they provide can, or should, be united (see, e.g., Dupré 1993; Mitchell 2003; Horst 2016).

In an attempt to offer some clarity and coherence, Gärdenfors and Stephens (2018) have argued that there are three nested basic forms of knowledge: procedural knowledge-how, conceptual knowledge-what, and propositional knowledge-that. The tri-partite knowledge-account is grounded in cognitive neuroscience where the three forms of knowledge are mapped to procedural, semantic, and episodic memory respectively. While there is an extensive and on-going epistemological discussion concerning the traditional forms knowledge-how and knowledge-that (see, e.g., Ryle 1949; Stanley 2011; Fantl 2016), a lot remains to be explored regarding the form knowledge-what, which Gärdenfors and Stephens argue is generated by inductive reasoning.

Moreover, in encouragement of a multi-disciplinary and multi-level development of our understanding of knowledge, cognition and behavior (see, e.g., Frank and Badre 2015), it can be pointed out that:

> [T]he neurosciences are reshaping the landscape of the behavioral sciences, and the behavioral sciences are of increasing importance to the neurosciences, especially for the rapidly expanding investigations into the highest level functions of the brain. (Berntson and Cacioppo 2009, p. xi)

This article attempts to broaden the proposed knowledge-account and our understanding of knowledge-what by investigating two issues. First, the prospect of integrating the knowledge-account with models from two other scientific perspectives (cognitive ethology and cognitive psychology) on higher levels of analysis will be explored. If successful, such integration would increase the knowledge-account's plausibility. By encompassing three levels of analysis, it would present a naturalistic framework arguably fairly close to a traditional epistemological outlook. Second, the link between the knowledge-form knowledge-what and *categorization* will be considered. I will loosely follow a prototype theoretical interpretation and view categorizations as natural cognitive phenomena where organisms try to acquire as much information as possible of the surrounding structured world, while minimizing their energy-expenditure (see, e.g., Rosch 1975a, b). According to such an interpretation, objects in a category are compared in relation to how representable they are, and the most representable object is seen as a prototype. Other objects can then be compared in relation to how similar they are to the prototype (see, e.g., Gärdenfors 2000, p. 84). This inquiry diverges from Gärdenfors and Stephens' discussion, which centers on the specific role of inductive inferences, and can thus be seen as a complementary development of their account.

After this short introduction Sect. 4.2 will give an outline of Gärdenfors and Stephens' knowledge-account grounded in cognitive neuroscience. Section 4.3 then investigates knowledge from the perspective of cognitive ethology, and the possibility of integrating the cognitive ethological account with the neuroscientific account. Section 4.4 continues by inquiring into how a cognitive psychological account can be integrated with both former accounts, and in Sect. 4.5 it is lastly argued that conceptual knowledge-what and categorizations can be fruitfully modeled within a conceptual spaces framework (Gärdenfors 1990, 2000, 2014).

4.2 Cognitive Neuroscience: Knowledge and Memory

Even though there are various different models and theories pertinent to understand knowledge from a neuroscientific proximate perspective, Gärdenfors and Stephens (2018) single out and use Tulving's (1985; see also 1972) seminal account of memory and consciousness. This is, arguably, a reasonable basis since Tulving's account has been extremely influential and is often used as a starting-point in neuroscientific research even by those who ultimately deviate from it. Knowledge, from a neuroscientific perspective, is thought to have its foundation in long-term memory (LTM), and Tulving divides LTM into three nested parts, illustrated in Fig. 4.1: procedural memory, semantic memory, and episodic memory, where '[...] procedural memory entails semantic memory as a specialized subcategory, and in which semantic memory, in turn, entails episodic memory as a specialized subcategory.' (Tulving 1985, pp. 2–3, italics removed; see also Fletcher et al. 1999; Goel and Dolan 2000; Kan et al. 2009; Barrett 2015; Kim 2016). Tulving argues that:

> Procedural memory [...] is concerned with how things are done – with the acquisition, retention, and utilization of perceptual, cognitive, and motor skills. Semantic memory – also called generic [...] or categorical memory [...] – has to do with the symbolically representable knowledge that organisms possess about the world. Episodic memory mediates the *remembering* of personally experienced events [...]. (Tulving 1985, p. 2)

With this partitioning as an underpinning, and trailing the neuroscientific canon, procedural knowledge-how (the knowledge of *how* to ride a bike – an ability) readily

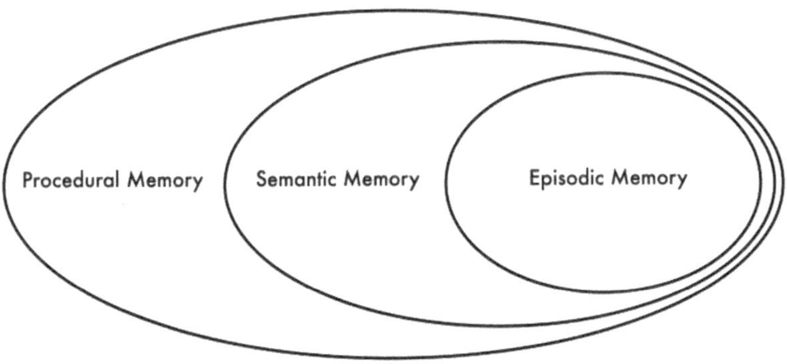

Fig. 4.1 Tulving's nested account of the LTM. Procedural memory entails semantic memory as a specialized subcategory, and semantic memory entails episodic memory as a specialized subcategory

maps to non-declarative procedural memory. This form of memory governs actions while it to a large extent is automatic and non-conscious. Through repetition we can learn, but we do it without being able to put all aspects of this knowledge into words. To use the above-mentioned example, we might be able to describe – in broad terms – what one should think about when learning how to ride a bike. But these instructions will not be enough to master the complicated motoric patterns necessary to execute the ability. This form of knowledge and learning-process instead demands practice. Procedural memory relies on the complex and interconnected performance of perceptual and motor pathways, involving, for example, the basal ganglia, neocortex, cerebellum, striatum, and the premotor- and primary motor cortex (see, e.g., Kandel et al. 2013). Many animals are endowed with procedural memory and are capable of procedural knowledge-how (Tulving 2002).

Semantic memory governs 'an individual's store of knowledge about the world. The content of semantic memory is abstracted from actual experience and is therefore said to be conceptual, that is, generalized and without reference to any specific experience.' (Binder and Desai 2011, p. 527). Semantic memory is crucial for numerous animals navigating a complex world (Roberts 2016). Moreover, an agent's ability to contemplate concepts and their relations, to perform inductive inferences and, as I want to emphasize, to *categorize* are all linked to semantic memory:

> Categorization is fundamental to understanding and using the concepts in semantic memory, since this process helps organize our knowledge and relate a test object to other known objects in the world. Categorization also allows us to engage in activities such as understanding unfamiliar objects and learning about novel objects. (Grossman et al. 2002b, p. 1549)

Gärdenfors and Stephens (2018) map conceptual knowledge-what (the knowledge of *what* a category consists in: dogs *characteristically* have four legs) to semantic memory. Similar formulations are indeed already in use in neuroscientific discussions:

> Thus humans use conceptual knowledge for much more than merely interacting with objects. All of human culture, including science, literature, social institutions, religion, and art, is constructed from conceptual knowledge. We do not reason, plan the future or remember the past without conceptual content – all of these activities depend on activation of concepts stored in semantic memory. (Binder and Desai 2011, p. 527)

Furthermore, several fMRI studies link the neural correlations of semantic encoding and semantic processing to '[. . .] many cognitive tasks, from perception, categorization, to explicit reasoning in problem-solving and decision-making.' (Goel and Dolan 2000, p. 110). In fact, many findings directly link semantic memory and categorization. Although, some discrepancies in neural activation is to be expected depending on, among other factors, variability regarding which aspects is in focus and regarding how stimulus is presented to test subjects (see, e.g., Grossman et al. 2002a). For example, Yee et al. (2014) explicitly relate conceptual knowledge to semantic memory and claim that such knowledge is distributed over many brain regions, which makes it flexible and able to handle varying contexts.

Semantic memory relies on associative pathways, involving, amongst other areas, the prefrontal cortex, the lateral-, ventral- and medial temporal cortex, basal ganglia, and hippocampus (see, e.g., Kandel et al. 2013):

> Semantic knowledge is stored in distinct association cortices and retrieval depends on the prefrontal cortex. [...] Semantic knowledge is distinguished from episodic knowledge in that it is typically not associated with the context in which the information was acquired. It is stored in a distributed manner in the neocortex, including the lateral and ventral temporal lobes. (Kandel et al. 2013, pp. 1449–1450)

Lastly, propositional knowledge-that (the knowledge *that* Stockholm is the capital of Sweden) maps to declarative episodic memory, governing factual remembrances and a sense of time – thereby playing a large part in how agents plan for the future:

> Memory for specific experiences is called episodic memory, although the content of episodic memory depends heavily on retrieval of conceptual knowledge. Remembering, for example, that one had coffee and eggs for breakfast requires retrieval of the concepts of coffee, eggs and breakfast. Episodic memory might be more properly seen as a particular kind of knowledge manipulation that creates spatial-temporal configurations of object and event concepts. (Binder and Desai 2011, p. 527)

Gärdenfors and Stephens (2018), ascribe facts to episodic memory (propositional knowledge-that) rather than to semantic memory (conceptual knowledge-what) – an interpretation somewhat similar to how for example Renoult et al. (2016) view 'autobiographical facts' as grounded in episodic memory. Episodic memory is crucially involved in self-awareness and first-person phenomenology. Since it according to Tulving's account is an evolutionarily later specialized subcategory, as shown in Fig. 4.1, it is largely dependent on semantic memory:

> Episodic memory refers to a complex and multifaceted process which enables the retrieval of richly detailed evocative memories from the past. In contrast, semantic memory is conceptualized as the retrieval of general conceptual knowledge divested of a specific spatiotemporal context. [... T]he available evidence [...] converges to highlight the pivotal role of semantic memory in providing schemas and meaning whether one is engaged in autobiographical retrieval for the past, or indeed, is endeavoring to construct a plausible scenario of an event in the future. It therefore seems plausible to contend that semantic processing may underlie most, if not all, forms of episodic memory, irrespective of temporal condition. (Irish and Piguet 2013, p. 1)

Episodic memory relies on attentional pathways, involving, for example, the prefrontal cortex, and the ventral-fronto- and medial temporal cortex (see, e.g., Kandel et al. 2013).

Episodic memory is conventionally considered uniquely human although there is increasing evidence indicating that animals – primarily rats, corvids, and great apes – have some form of episodic memory. For example Panoz-Brown et al. (2016, p. 2821; see also Roberts 2016) argue that '[...] rats remember multiple unique events and the contexts in which these events occurred using episodic memory and support the view that rats may be used to model fundamental aspects of human cognition.' Clayton et al. (2001, p. 1483) contend that '[...] jays form integrated memories for the location, content and time of caching. This memory capability fulfills Tulving's behavioural criteria for episodic memory and is thus termed "episodic-like".' Rilling et al. (2007, p. 17149) describe how their '[...] results raise the possibility that the resting state of chimpanzees involves emotionally laden episodic memory retrieval and some level of mental self-projection, albeit in the absence of language and conceptual processing.' As a last example, Allen and Fortin (2013, p. 10380) even claim that '[...] core properties of episodic memory are present across mammals, as well as in a number of bird species.'

Tulving (2005) discusses the issue of episodic memory in animals and points out that:

> It depends partly on what one means by episodic memory, partly on the kinds of evidence one considers, and partly on how one interprets the evidence. When episodic memory is defined loosely as 'memory for (specific) past events,' then the standard commonsense answer is that of course animals have it. (Tulving 2005, p. 35)

However, Tulving highlights the importance of a less anthropomorphic perspective than this 'commonsense' understanding. Focusing on mental time travel, which is an essential aspect of episodic memory in humans and a distinguishing trait, he argues that:

> [...] only human beings possess "autonoetic" episodic memory and the ability to mentally travel into the past and into the future, and that in that sense they are unique. (Tulving 2005, p. 4)

The issue might be impossible to conclusively settle, since there are valid arguments for a variety of interpretations that ultimately hinge on how one *choose* to interpret the relevant terms, theories and evidence. Nevertheless, even if one accepts that animals other than humans can have episodic memories; it is to a significantly lesser degree. This fits with the view that episodic memory (propositional knowledge) is evolutionarily subsequent to the two other forms of LTM (Tulving 1985, 2002, 2005).

As previously mentioned, a way to increase the plausibility of the above-described knowledge-account is to investigate whether it is possible to integrate with models – from other sciences – on other levels of analysis. Since '[t]he neural basis of behavior cannot be properly characterized without first allowing for independent detailed study of the behavior itself [...]' (Krakauer et al. 2017, p. 488), the next section will explore the possibility of such integration by using Kornblith's (2002) analysis and account of knowledge from cognitive ethology (see also, e.g., Mitchell 2003; Cellucci 2017).

4.3 Cognitive Ethology: Evolution and Reliability

'The biological study of animal behavior, including its phenomenological, causal, ontogenetic, and evolutionary aspects, is a discipline known as ethology' (Anderson and Perona 2014, p. 18). Ethology investigates animal behavior concentrating on natural environmental settings. Moreover, there is an ongoing discussion if such behavior is intentional – and if so, to what degree (see, e.g., Allen and Bekoff 1995;

Wynne 2007; Shettleworth 2010). From an ultimate perspective, knowledge can be seen as the result of a phylogenetic and genotypic adaptive (functional) process, which shapes the cognitive faculties of agents (see, e.g., Plotkin 1993; Avital and Jablonka 2000):

> What is actually meant is that knowledge is a complex set of relationships between genes and past selection pressures, between genetically guided developmental pathways and the conditions under which development occurs, and between a part of the consequent phenotypic organization and specific features of environmental order. (Plotkin 1993, p. 228)

Our cognitive faculties are the result of evolutionary processes that has formed our sense organs and cognitive architecture. So, our evolutionarily molded cognitive faculties enable, as well as constrain, what we know (Plotkin 1993, p. 162).

In addition to these innate features, agents can acquire knowledge by *learning*, an ontogenetic aspect '[. . .] indicating processes by which the individual, thanks to phenotypic modifications, accommodates to novel circumstances in the course of its life.' (Serrelli and Rossi 2009, p. 18). In connection to ethology, implicit learning and implicit memory are central '[. . . involving] a wide variety of brain regions, most often cortical areas that support the specific perceptual, conceptual, or motor systems recruited to process a stimulus or perform a task.' (Kandel et al. 2013, p. 1459). Implicit learning splits into non-associative and associative learning, where non-associative learning includes responses to repeatedly encountered stimulus, in the form of habituation, where an agent's response diminishes by repeated exposure to a stimulus, and sensitization, where exposure strengthens a response. Associative learning involves how agents learn to link (associate) different stimuli to each other, in the form of conditioning by stimulus, response, and grasped relationships (see, e.g., Kandel et al. 2013).

Non-associative and associative learning thus match procedural respectively semantic memory, and, even though the focus is on particular brain systems rather than on implicit memory generally, for example Ullman (2016) argues that:

> Procedural memory involves a network of interconnected brain structures rooted in frontal/basal-ganglia circuits, including frontal premotor and related regions, particularly BA 6 and BA 44. [. . .] This circuitry underlies the implicit (nonconscious) learning and processing of a wide range of perceptual- motor and cognitive skills, tasks, and functions [. . .] including navigation, sequences, rules, and categories. (Ullman 2016, p. 956)

Illuminating the cognitive ethological position, Kornblith (2002, see also 1993) offers a fruitful discussion about 'fitness' and how animals that have knowledge about their changing environment better survive and thrive.[1] In a more traditional epistemological terminology, he points out that cognitive ethology provides:

> [. . .] a large literature on animal cognition, and [how] workers in this field typically speak of animals knowing a great many things. They see animal knowledge as a legitimate object of study, a phenomenon with a good deal of theoretical integrity to it. Knowledge, as it is portrayed in this literature, does causal and explanatory work. (Kornblith 2002, pp. 28–29)

[1] For a critique of Kornblith's position see for example Bermúdez (2006).

According to Kornblith's interpretation, cognitive ethology supports a reliabilist account of knowledge where knowledge should be seen as demanding reliably produced true beliefs (*RTB*)[2]:

> [...] I will argue that the kind of knowledge that philosophers have talked about all along just is the kind of knowledge that cognitive ethologists are currently studying. Knowledge explains the possibility of successful behavior in an environment, which in turn explains fitness. [... W]e must appeal to a capacity to recognize features of the environment, and thus the true beliefs that [... someone] acquire will be the product of a stable capacity for the production of true beliefs. The resulting true beliefs are not merely accidentally true; they are produced by a cognitive capacity that is attuned to its environment. In a word, the beliefs are reliably produced. The concept of knowledge which is of interest here thus requires reliably produced true belief. (Kornblith 2002, pp. 29–30, 57–58)

As reliabilism is generally coupled with externalist forms of justification such as *truth-connectivity* and *reliability* where an agent does not need to have cognitive access to her beliefs, it fits well with the description of the nonconscious non-associative and associative learning (see, e.g., Kandel et al. 2013; Ullman 2016).[3] An integration of the cognitive ethological reliabilist account and the cognitive neuroscientific account is accordingly possible by focusing on the two evolutionarily prior forms of memory and knowledge – procedural memory (procedural knowledge-how) and semantic memory (conceptual knowledge-what).

4.4 Cognitive Psychology: Intuition and Deliberation

Cognitive psychology investigates how human mental processes, including knowledge, are connected to behavior, using both bottom-up and top-down methods.

On a psychological level of analysis, implicit memory, implicit learning, and non-associative learning are all seen as being linked to procedural memory (procedural knowledge). In various forms of behaviorism these concepts have been investigated with a focus on reinforcement and punishment. However, in many theories, explicit memory and explicit learning take a central place, governing rule learning, awareness, and active remembrance of facts, being linked to episodic memory (propositional knowledge) (Kandel et al. 2013):

> [Explicit memory] is the deliberate or conscious retrieval of previous experiences as well as conscious recall of factual knowledge about people, places, and things. [...] Explicit memory is highly flexible; multiple pieces of information can be associated under different circumstances. (Kandel et al. 2013, p. 1446)

---

[2]Kornblith argues that cognitive ethology 'gives us the *only* viable account of what knowledge is.' (Kornblith 2002, p. 135, my italics). However, he does not motivate this restriction in a convincing way – pointed out by for example Kusch (2005) – and so this aspect of Kornblith's otherwise fruitful ideas will not be heeded here.

[3]Episodic memory (propositional knowledge) governing self-awareness and first-person phenomenology, on the other hand, is more naturally linked to internalism and forms of justification such as *rationality* and *cognitive access* (Tulving 2005; Alston 2005).

As an in-between, semantic memory (conceptual knowledge) is involved in both implicit and explicit memory, being linked to associative learning, pattern recognition, categorization, and prototype-matching. Regarding conceptual knowledge and categorization, for example Csibra and Gergely (2006) inquire into how teaching, and learning from teaching, should be viewed as a key adaptation for the transfer of knowledge between humans (see also Gärdenfors and Högberg 2017; Gergely et al. 2007). To facilitate social learning and teaching, they highlight how pedagogy offers a possibility to transfer generalizable knowledge, instead of just factual information, from a (active) teacher to a learner. Such generalizable knowledge does not only pertain to a specific situation but can be applied in many different contexts, which is essential for the ability to categorize. Csibra and Gergely (2009) develops their thoughts on generalizable knowledge:

> If I point at two aeroplanes and tell you that 'aeroplanes fly', what you learn is not restricted to the particular aeroplanes you see or to the present context, but will provide you generic knowledge about the kind of artefact these planes belong to that is generalizable to other members of the category and to variable contexts. Moreover, the transmission of such generic knowledge is not restricted to linguistic communication. If I show you by manual demonstration how to open a milk carton, what you will learn is how to open that kind of container (i.e. you acquire kind-generalizable knowledge from a single manifestation). In such cases, the observer does not need to rely on statistical procedures to extract the relevant information to be generalized because this is selectively manifested to her by the communicative demonstration. (Csibra and Gergely 2009, p. 148)

This type of generic generalizable knowledge, associated with categorization, seems reasonable to view as conceptual knowledge-what. Gärdenfors and Högberg point out that 'communicating concepts' is an evolutionarily prior form of teaching to 'explaining relationships between concepts' (Gärdenfors and Högberg 2017, pp. 193–195). According to Gärdenfors and Högberg, 'communicating concepts' at its core involves pattern-recognition, linking it to categorization and conceptual knowledge-what. 'Explaining relationships between concepts', on the other hand, involves teaching of facts and symbolic language making it more readily linked to propositional knowledge-that (Gärdenfors and Högberg 2017, pp. 193–195).

A well-established position in cognitive psychology is that of the *dual process framework* (see, e.g., Lizardo et al. 2016). Specifically the (default-interventionist) *dual process theory* has been prominent, which divides mental processing into one unconscious implicit and one conscious explicit reasoning system (see, e.g., Bago and De Neys 2017; Lizardo et al. 2016; Huberdeau et al. 2015; Sloman 1993, 2014; Evans and Stanovich 2013; Kahneman 2011; Rugg and Curran 2007):

- *System 1* operates automatically and quickly, with little or no effort and no sense of voluntary control. (Kahneman 2011, p. 20)
- *System 2* allocates attention to the effortful mental activities that demand it, including complex computations. The operations of System 2 are often associated with the subjective experience of agency, choice, and concentration. (Kahneman 2011, p. 21)

System 1 (or Type 1) can be described as intuitive and heuristic whereas System 2 (or Type 2) is deliberate and analytical, where the slow analytical process tries to inhibit the faster intuitive process.

There are a number of alternative theories arguing that cognition should be seen as consisting of a single process, as well as theories arguing for the possibility of parallel additional and/or more fine-grained systems (see, e.g., Bago and De Neys 2017; Rugg and Curran 2007). But I will follow Smith and DeCoster (2000, p. 110) who argue that 'numerous models of dual-processing modes can be integrated and interpreted in terms of the properties of two underlying memory systems and that this integration will lead to new insights and new predictions in several substantive areas of psychology.' (see also, e.g., Goel et al. 2000; Goel and Dolan 2003):

> The architecture that supports the interaction between systems has been hinted at in the cognitive neuroscience literature. Anatomically, the brain includes multiple parallel frontal corticobasal ganglia loops [. . .]. The interactions among these loops can be interpreted as a set of gating mechanisms [. . .]. My proposal is that one such loop is the intuitive loop, though it is best characterized as jointly intuitive and affective. Deliberation, in contrast, involves a more anterior prefrontal corticobasal ganglia loop. One critical function of deliberation is to serve to gate or at least modulate the intuitive–affective loop. (Sloman 2014, p. 75)

'System 1 is generally described as a form of universal cognition shared between humans and animals [. . . and] System 2 is believed to have evolved much more recently and is thought by most theorists to be uniquely human.' (Evans 2003, p. 454; see also Evans and Stanovich 2013, p. 225):

> Although rudimentary forms of higher order control can be observed in mammals and other animals [. . .], the controlled processing in which they can engage is very limited by comparison with humans, who have unique facilities for language and meta-representation as well as greatly enlarged frontal lobes [. . .]. We are in agreement that the facility for Type 2 thinking became uniquely developed in human beings, effectively forming a new mind [. . .], which coexists with an older mind based on instincts and associative learning and gives humans the distinctive forms of cognition that define the species [. . .]. (Evans and Stanovich 2013, p. 236)

System 1 is thus arguably compatible with the aforementioned *RTB*-account from cognitive ethology, and the two evolutionarily earlier memory forms (and knowledge forms) from cognitive neuroscience since '[t]he capabilities of System 1 include innate skills that we share with other animals.' (Kahneman 2011, p. 21):

> System 1 is old in evolutionary terms and shared with other animals: it comprises a set of autonomous subsystems that include both innate input modules and domain-specific knowledge acquired by a domain-general learning mechanism. System 2 is evolutionarily recent and distinctively human: it permits abstract reasoning and hypothetical thinking, but is constrained by working memory capacity and correlated with measures of general intelligence. (Evans 2003, p. 454)

In other words can System 1, on a cognitive psychological level of analysis, be mapped to procedural memory (procedural knowledge) and semantic memory (conceptual knowledge), on a cognitive neuroscientific level of analysis, and to *RTB*, on a cognitive ethological level of analysis. System 1 is thus most naturally linked to externalist justification – even though semantic memory (conceptual knowledge) can be seen as an 'in-between,' containing both externalist and internalist elements.

By adding episodic memory (propositional knowledge), on the neuroscientific level of analysis, System 2, on the cognitive psychological level of analysis, can be illuminated. System 2 is viewed as 'the conscious, reasoning self that has beliefs, makes choices, and decides what to think about and what to do' (Kahneman 2011, p. 21). Episodic memory, governs conscious and active reflection where we have cognitive access to our beliefs, on the neuroscientific level of analysis. It thus makes it possible to account for internalist justification and 'the subjective experience of agency, choice, and concentration' (Kahneman 2011, p. 21), on the cognitive psychological level of analysis, which is needed to fully explain human cognition. The three memory systems, on the neuroscientific level of analysis, can hence explain both System 1 and System 2 on the cognitive psychological level of analysis. In support of such integration for example Lizardo et al. (2016) explicitly connect 'know how' and non-declarative representation to System 1, whereas 'know that' and declarative representation is connected to System 2:

> [. . . M]emory is divided into two main types, most commonly referred to as "declarative" and "nondeclarative" memory. Declarative memory (Type II) consists of consciously accessible memories of facts, symbols, and events, while nondeclarative memory (Type I) consists of relatively less accessible procedural knowledge, habits, and dispositions. The two kinds of memory are sometimes distinguished as "knowing that" and "knowing-how" [. . .], or "explicit" and "implicit" memory [. . .]. (Lizardo et al. 2016, section 3.2)

The discussion points in the direction of compatibility and a plausible integration of the models on the presented three levels of analysis.

4.5 Conceptual Spaces: Knowledge-What and Categorization

The conceptual spaces framework has been presented and developed by Gärdenfors as a complementary alternative to the conventional symbolic and subconceptual forms of representation (Gärdenfors 1990, 2000, 2014). It postulates geometrical structures, where 'phenomenal' quality dimensions are grouped into domains. Observations of objects can then, in accordance with their properties, be positioned in a dimensional region. Properties can thereafter be compared in regard of their relation, where relative proximity represents degree of 'similarity.' To fruitfully analyze properties, categories, and their relations, Gärdenfors proposes two definitional criteria that provides spatial structure:

> *Criterion P*: A *natural property* is a convex region of a domain in a conceptual space, and;
> *Criterion C*: A *natural concept [or category]* is represented as a set of convex regions in a number of domains together with an assignment of salience weights to the domains and information about how the regions in different domains are correlated.

The convexity of criterion *P* is thought to capture that if two objects, exemplifying a property, are located in a particular domain, then objects positioned *between* those objects will also exemplify that same property. Criterion *C* highlights that natural concepts and categories are based on one *or more* domains – a distinction that is lost in the traditional language-focused approach. Conceptual spaces thus offer the ability to add and adjust dimensions in a domain, making it possible to elucidate how they are similar and/or connected. Furthermore, conceptual spaces make it possible to clarify and explain category-formation and learning. So, by utilizing conceptual spaces and criteria *P* and *C* it is thus possible to model categorizations and the knowledge linked to them; i.e. conceptual knowledge-what (see, e.g., Gärdenfors 2000, 2014; see also Douven et al. 2013; Decock et al. 2013).

Focusing on knowledge-what and categorization, Gärdenfors and Williams (2001) specifically address how categorizations efficiently can be modeled with conceptual spaces. Even though their focus is on artificial intelligence, the framework has the ability to clearly show prototypes, independent dimensions, and similarity. They point out that '[t]here is a wealth of psychological data supporting the existence of prototypes and their key role in categorization' (Gärdenfors and Williams 2001, p. 387):

> In summary the key findings from psychological studies of categorization are (i) similarity judgments play a fundamental role in categorization and they are context sensitive, (ii) the degree of similarity is judged with respect to a reference object/region such as a prototype, (iii) category membership can be graded (discrete membership, if and when it exists, is considered to be a special case), and (iv) the psychophysical relationship between the stimulus and the response depends on the underlying categorization. (Gärdenfors and Williams 2001, p. 387)

I regard it to ultimately be up to any theoretician to investigate those domains and quality dimensions that are found to be of interest. But reconnecting to the above discussion about our evolutionarily molded cognitive faculties; there are some innate, natural, domains and quality dimensions for humans, and '[o]ur quality dimensions are what they are because they have been selected to fit the surrounding world.' (Gärdenfors 2000, p. 82).[4] Taken together, this strongly indicates that conceptual spaces are apt for investigating categorization and conceptual knowledge-what – the knowledge of what a category characteristically consists in.

4.6 Concluding Remarks

An integration of the neuroscientifically grounded knowledge-account with accounts from cognitive ethology and cognitive psychology has been shown to be plausible. Procedural and semantic memory, on a neuroscientific level of analysis, match an ethological reliabilist account, as well as System 1 from the psychological dual process theory. By adding episodic memory, on the neuroscientific level of

[4]More or less similar domains and quality dimensions can also be found for other animals (see, e.g., Lorenz 1973; for an illuminating classic discussion see also Nagel 1974).

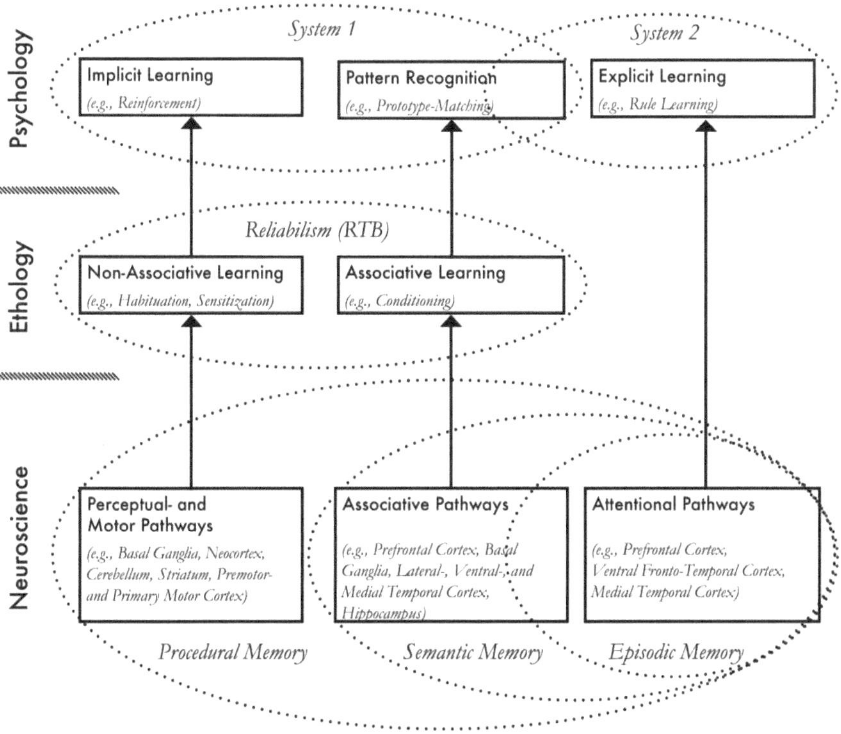

Fig. 4.2 Knowledge seen from a neuroscientific, ethological, and psychological level of analysis. Dotted lines indicate knowledge-categories; boxes outline examples of more detailed content descriptions, and; arrows show hierarchical mappings

analysis, System 2 on the psychological level can be accounted for. The article's integrative view is illustrated in Fig. 4.2.

This three-level naturalistic epistemological framework, linking conceptual knowledge-what to categorizations – fruitfully modeled within a conceptual spaces framework – promises interesting ramifications. On one hand it might fill a deleterious role and exert a dissolving influence on traditional epistemological problems and paradoxes. This is so since it moves a lot of focus away from propositional knowledge-that, which for a long time has had the center stage in epistemology, to conceptual knowledge-what. Moreover, it should impact discussions regarding, for example, reductionism since all three memory forms are considered important in their own right, which might be viewed as an argument against reduction. Importantly, *if* there is a reduction to be made it should be from propositional knowledge-that to conceptual knowledge-what and/or procedural knowledge-how, or from conceptual knowledge-what to procedural knowledge-how – *not* the other way around. On the other hand this nested take on naturalistic epistemology also offers a way to discover more and new scientifically grounded details regarding knowledge on other levels of analysis.

Acknowledgments I have had the great pleasure and privilege of investigating and discussing these topics with Peter Gärdenfors, and I am very grateful to Peter for sharing his vast knowledge, his eye for detail, and his positive energy. I would also like to thank Mauri Kaipainen for his generous and insightful remarks. Thanks to Trond Arild Tjøstheim for inspiring discussions. Finally I would like to thank my anonymous reviewers for comments.

References

Allen, C., & Bekoff, M. (1995). Cognitive ethology and the intentionality of animal behaviour. *Mind & Language, 10*(4), 313–328.

Allen, T. A., & Fortin, N. J. (2013). The evolution of episodic memory. *Proceedings of the National Academy of Sciences of the United States of America, 110*(Supplement 2), 10379–10386.

Alston, W. P. (2005). *Beyond "justification": Dimensions of epistemic evaluation*. Ithaca: Cornell University Press.

Anderson, D. J., & Perona, P. (2014). Toward a science of computational ethology. *Neuron, 84*(1), 18–31.

Avital, E., & Jablonka, E. (2000). *Animal traditions: Behavioural inheritance in evolution*. Cambridge: Cambridge University Press.

Bago, B., & De Neys, W. (2017). Fast logic? Examining the time course assumption of dual process theory. *Cognition, 158*, 90–109.

Barrett, H. C. (2015). *The shape of thought: How mental adaptations evolve*. Oxford: Oxford University Press.

Bermúdez, J. L. (2006). Knowledge, naturalism, and cognitive ethology: Kornblith's *Knowledge and its place in nature*. *Philosophical Studies, 127*(2), 299–316.

Berntson, G. G., & Cacioppo, J. T. (2009). Preface. In G. G. Berntson & J. T. Cacioppo (Eds.), *Handbook of neuroscience for the behavioral science* (pp. xi–xii). New York: Wiley.

Binder, J. R., & Desai, R. H. (2011). The neurobiology of semantic memory. *Trends in Cognitive Sciences, 15*(11), 527–536.

Cellucci, C. (2017). *Rethinking knowledge: The heuristic view* (European studies in philosophy of science, Vol. 4). Springer.

Clayton, N. S., Griffiths, D. P., Emery, N. J., & Dickinson, A. (2001). Elements of episodic-like memory in animals. *Philosophical Transactions of the Royal Society, B: Biological Sciences, 356*(1413), 1483–1491.

Csibra, G., & Gergely, G. (2006). Social learning and social cognition: The case for pedagogy. In Y. Munakata & M. H. Johnson (Eds.), *Processes of change in brain and cognitive development. Attention and performance* (Vol. XXI, pp. 249–274). Oxford: Oxford University Press.

Csibra, G., & Gergely, G. (2009). Natural pedagogy. *Trends in Cognitive Sciences, 13*(4), 148–153.

Decock, L., Dietz, R., & Douven, I. (2013). Modelling comparative concepts in conceptual spaces. In Y. Motomura, A. Butler, & D. Bekki (Eds.), *New frontiers in artificial intelligence* (pp. 69–86). Heidelberg: Springer.

Douven, I., Decock, L., Dietz, R., & Égré, P. (2013). Vagueness: A conceptual spaces approach. *Journal of Philosophical Logic, 42*(1), 137–160.

Dupré, J. (1993). *The disorder of things: Metaphysical foundations of the disunity of science*. Cambridge, MA: Harvard University Press.

Evans, J. S. B. (2003). In two minds: Dual-process accounts of reasoning. *Trends in Cognitive Sciences, 7*(10), 454–459.

Evans, J. S. B., & Stanovich, K. E. (2013). Dual-process theories of higher cognition: Advancing the debate. *Perspectives on Psychological Science, 8*(3), 223–241.

Fantl, J. (2016). Knowledge how. In E. N. Zalta (Ed.), *The Stanford encyclopedia of philosophy* (Spring 2016 Edition). https://plato.stanford.edu/archives/spr2016/entries/knowledge-how/

Fletcher, P. C., Büchel, C., Josephs, O., Friston, K., & Dolan, R. J. (1999). Learning-related neuronal responses in prefrontal cortex studied with functional neuroimaging. *Cerebral Cortex, 9*(2), 168–178.

Frank, M. J., & Badre, D. (2015). How cognitive theory guides neuroscience. *Cognition, 135*, 14–20.

Gärdenfors, P. (1990). Induction, conceptual spaces and AI. *Philosophy of Science, 57*(1), 78–95.

Gärdenfors, P. (2000). *Conceptual spaces: The geometry of thought*. Cambridge, MA: Bradford Books/MIT Press.

Gärdenfors, P. (2014). *The geometry of meaning: Semantics based on conceptual spaces*. Cambridge, MA: MIT Press.

Gärdenfors, P., & Högberg, A. (2017). The archaeology of teaching and the evolution of *Homo docens*. *Current Anthropology, 58*(2), 188–208.

Gärdenfors, P., & Stephens, A. (2018). Induction and knowledge-what. *European Journal for Philosophy of Science, 8*(3), 471–491.

Gärdenfors, P., & Williams, M. A. (2001). Reasoning about categories in conceptual spaces. In *Proceedings of the Fourteenth International Joint Conference of Artificial Intelligence* (pp. 385–392). Morgan Kaufmann Publishers.

Gergely, G., Egyed, K., & Király, I. (2007). On pedagogy. *Developmental Science, 10*(1), 139–146.

Goel, V., & Dolan, R. J. (2000). Anatomical segregation of component processes in an inductive interference task. *Journal of Cognitive Neuroscience, 12*(1), 110–119.

Goel, V., & Dolan, R. J. (2003). Explaining modulation of reasoning by belief. *Cognition, 87*(1), B11–B22.

Goel, V., Buchel, C., Frith, C., & Dolan, R. J. (2000). Dissociation of mechanisms underlying syllogistic reasoning. *NeuroImage, 12*(5), 504–514.

Goldman, A., & Beddor, B. (2016). Reliabilist epistemology. In E. N. Zalta (Ed.), *The Stanford encyclopedia of philosophy* (Winter 2016 Edition). https://plato.stanford.edu/archives/win2016/entries/reliabilism/

Grossman, M., Koenig, P., DeVita, C., Glosser, G., Alsop, D., Detre, J., & Gee, J. (2002a). The neural basis for category-specific knowledge: An fMRI study. *NeuroImage, 15*(4), 936–948.

Grossman, M., Smith, E. E., Koenig, P., Glosser, G., DeVita, C., Moore, P., & McMillan, C. (2002b). The neural basis for categorization in semantic memory. *NeuroImage, 17*(3), 1549–1561.

Horst, S. (2016). *Cognitive pluralism*. Cambridge: The MIT Press.

Huberdeau, D. M., Krakauer, J. W., & Haith, A. M. (2015). Dual-process decomposition in human sensorimotor adaptation. *Current Opinion in Neurobiology, 33*, 71–77.

Irish, M., & Piguet, O. (2013). The pivotal role of semantic memory in remembering the past and imagining the future. *Frontiers in Behavioral Neuroscience, 7*(27), 1–11.

Kahneman, D. (2011). *Thinking fast and slow*. New York: Farrar, Straus and Giroux.

Kan, I. P., Alexander, M. P., & Verfaellie, M. (2009). Contribution of prior semantic knowledge to new episodic learning in amnesia. *Journal of Cognitive Neuroscience, 21*(5), 938–944.

Kandel, E. R., Schwartz, J. H., Jessell, T. M., Siegelbaum, S. A., & Hudspeth, A. J. (Eds.). (2013). *Principles of neural science* (5th ed.). New York: McGraw-Hill, Health Professions Division.

Kim, H. (2016). Default network activation during episodic and semantic memory retrieval: A selective meta-analytic comparison. *Neuropsychologia, 80*, 35–46.

Kornblith, H. (1993). *Inductive inference and its natural ground: An essay in naturalistic epistemology*. Cambridge: MIT Press.

Kornblith, H. (2002). *Knowledge and its place in nature*. Oxford: Oxford University Press.

Krakauer, J. W., Ghazanfar, A. A., Gomez-Marin, A., MacIver, M. A., & Poeppel, D. (2017). Neuroscience needs behavior: Correcting a reductionist bias. *Neuron, 93*(3), 480–490.

Kusch, M. (2005). Beliefs, kinds and rules: A comment on Kornblith's *Knowledge and its place in nature*. *Philosophy and Phenomenological Research, 71*(2), 411–419.

Lizardo, O., Mowry, R., Sepulvado, B., Stoltz, D. S., Taylor, M. A., Van Ness, J., & Wood, M. (2016). What are dual process models? Implications for cultural analysis in sociology. *Sociological Theory, 34*(4), 287–310.

Lorenz, K. (1973/1977). *Behind the mirror: A search for a natural history of human knowledge*. London: Methuen.
Markie, P. (2013). Rationalism vs. empiricism. In E. N. Zalta (Ed.), *The Stanford encyclopedia of philosophy* (Summer 2013 Edition). http://plato.stanford.edu/archives/sum2013/entries/rationalism-empiricism/
Mitchell, S. D. (2003). *Biological complexity and integrative pluralism*. Cambridge: Cambridge University Press.
Nagel, T. (1974). What is it like to be a bat? *The Philosophical Review, 83*(4), 435–450.
Panoz-Brown, D., Corbin, H. E., Dalecki, S. J., Gentry, M., Brotheridge, S., Sluka, C. M., Wu, J., & Crystal, J. D. (2016). Rats remember items in context using episodic memory. *Current Biology, 26*(20), 2821–2826.
Pappas, G. (2017). Internalist vs. externalist conceptions of epistemic justification. In E. N. Zalta (Ed.), *The Stanford encyclopedia of philosophy* (Fall 2017 Edition). https://plato.stanford.edu/archives/fall2017/entries/justep-intext/
Plotkin, H. C. (1993). *Darwin machines and the nature of knowledge*. Cambridge, MA: Harvard University Press.
Renoult, L., Tanguay, A., Beaudry, M., Tavakoli, P., Rabipour, S., Campbell, K., Moscovitch, M., Levine, B., & Davidson, P. S. (2016). Personal semantics: Is it distinct from episodic and semantic memory? An electrophysiological study of memory for autobiographical facts and repeated events in honor of Shlomo Bentin. *Neuropsychologia, 83*, 242–256.
Rilling, J. K., Barks, S. K., Parr, L. A., Preuss, T. M., Faber, T. L., Pagnoni, G., Bremer, D., & Votaw, J. R. (2007). A comparison of resting-state brain activity in humans and chimpanzees. *Proceedings of the National Academy of Sciences of the United States of America, 104*(43), 17146–17151.
Roberts, W. A. (2016). Episodic memory: Rats master multiple memories. *Current Biology, 26*(20), R920–R922.
Rosch, E. (1975a). Cognitive reference points. *Cognitive Psychology, 7*(4), 532–547.
Rosch, E. (1975b). Cognitive representations of semantic categories. *Journal of Experimental Psychology: General, 104*(3), 192–233.
Rugg, M. D., & Curran, T. (2007). Event-related potentials and recognition memory. *Trends in Cognitive Sciences, 11*(6), 251–257.
Ryle, G. (1949). *The concept of mind*. London: Hutchinson.
Samet, J., & Zaitchik, D. (2014). Innateness and contemporary theories of cognition. In E. N. Zalta (Ed.), *The Stanford encyclopedia of philosophy* (Fall 2014 Edition). http://plato.stanford.edu/archives/fall2014/entries/innateness-cognition/
Serrelli, E., & Rossi, F. M. (2009). *A conceptual taxonomy of adaptation in evolutionary biology*. Draft paper 4 september. Milano: University of Milano Bicocca.
Shettleworth, S. J. (2010). *Cognition, evolution, and behavior*. Oxford: Oxford University Press.
Sloman, S. A. (1993). Feature-based induction. *Cognitive Psychology, 25*(2), 231–280.
Sloman, S. A. (2014). Two systems of reasoning: An update. In J. W. Sherman, B. Gawronski, & Y. Trope (Eds.), *Dual-process theories of the social mind* (pp. 69–79). New York: Guilford Press.
Smith, E. R., & DeCoster, J. (2000). Dual-process models in social and cognitive psychology: Conceptual integration and links to underlying memory systems. *Personality and Social Psychology Review, 4*(2), 108–131.
Stanley, J. (2011). *Know how*. Oxford: Oxford University Press.
Tulving, E. (1972). Episodic and semantic memory. In E. Tulving & W. Donaldson (Eds.), *Organization of memory* (pp. 381–403). New York: Academic.
Tulving, E. (1985). Memory and consciousness. *Canadian Psychology, 26*(1), 1–12.
Tulving, E. (2002). Episodic memory: From mind to brain. *Annual Review of Psychology, 53*(1), 1–25.
Tulving, E. (2005). Episodic memory and autonoesis: Uniquely human. In H. Terrance & J. Metcalfe (Eds.), *The missing link in cognition: Origins of self-reflective consciousness* (pp. 3–56). New York: Oxford University Press.

Ullman, M. T. (2016). The declarative/procedural model: A neurobiological model of language learning, knowledge and use. In G. Hickok & S. L. Small (Eds.), *The neurobiology of language* (pp. 953–968). London: Academic.
Wynne, C. D. L. (2007). What are animals? Why anthropomorphism is still not a scientific approach to behavior. *Comparative Cognition and Behavior Reviews, 2*, 125–135.
Yee, E., Chrysikou, E. G., & Thompson-Schill, S. L. (2014). Semantic memory. In K. Ochsner & S. Kosslyn (Eds.), *The Oxford handbook of cognitive neuroscience: Volume 1, core topics* (pp. 353–374). Oxford: Oxford University Press.

Chapter 5
Convexity Is an Empirical Law in the Theory of Conceptual Spaces: Reply to Hernández-Conde

Peter Gärdenfors

Abstract This article is a rejoinder to Hernández-Conde's (Synthese 194(10):4011–4037, 2017) criticism of the convexity criterion in the theory of conceptual spaces. His arguments in general claim that the convexity criterion could be false and that it therefore is problematic for the theory. However, this is a misunderstanding since the convexity criterion is put forward as an empirically testable law for concepts. The long list of cases where the convexity criterion could be false that Hernández-Conde presents rather exhibits the rich empirical content of the criterion.

José Hernández-Conde (2017) criticizes the convexity requirement in the theory of conceptual spaces. His article presents a long list of cases where he argues that the convexity criterion does not have a "mandatory character". It contains some valid points, to which I return below. However, the article is based on a misunderstanding of the role of the convexity criterion.

Most of the arguments by Hernández-Conde have the following character:

(1) If convexity is a valid criterion, then X would follow.
(2) It *could be* that X is not true.
(3) Hence, the convexity criterion is not supported.

Hernández-Conde is perfectly right that in the cases he considers premises (1) and (2) are valid. What I believe is a misunderstanding of the status of the convexity criterion leads to the conclusion (3). Hernández-Conde repeatedly writes about the "mandatory character" of the convexity criterion, but it is not clear to me what it meant by that expression. It is important to note, however, that the convexity criterion is not proposed as something that necessarily holds of an application, but as an *empirical law* that generates testable predictions (Gärdenfors 2000, 2014). The convexity criterion is what furnishes the theory of conceptual spaces with most of its empirical content. The fact that the consequences of the convexity criterion

---

P. Gärdenfors (✉)
Lund University Cognitive Science, Department of Philosophy, LUX, Lund, Sweden
e-mail: Peter.Gardenfors@lucs.lu.se

M. Kaipainen et al. (eds.), *Conceptual Spaces: Elaborations and Applications*, Synthese Library 405, https://doi.org/10.1007/978-3-030-12800-5_5

could be false simply means that the criterion is testable. If Hernández-Conde had presented counterexamples to the consequences of the criterion, then it would have been problematic. But he does not do that.

The 'could be' argument is repeated throughout the article. I will not discuss all cases, but focus on a few. A typical example is section 4.3 where he argues that the convexity of the regions of the color domain is "no guarantee" that the convexity criterion will work in non-perceptual domains. True, but the criterion is put forward as a testable hypothesis for other domains.

Hernández-Conde (2017, p. 4016) writes that I am "strongly committed to the thesis that the shapes and boundaries of conceptual regions are produced by a Voronoi tessellation of the conceptual hyperspace" (this Hernández-Conde calls Thesis V). He notes that I never express this thesis openly, but he claims that I accept it tacitly. That is not true – for several reasons.

Firstly, for many basic dimensions – such as length, weight, age, temperature, and time – there exist no prototypes and, consequently, Voronoi tessellations are not definable for these dimensions. However, adjectives that are used to express properties determined by such dimensions ('long' vs. 'short', 'heavy' vs. 'light', etc) nevertheless represent convex regions. Hernández-Conde writes that "concepts show prototypical effects" (2017, p. 4017). Yes many do, but not the examples above. There are connections between conceptual spaces and prototype theory, but there is no "mutual dependence". For example, Hernández-Conde argues correctly that "[t]he only things that should be expected by a consistent prototype theorist is the star-shapedness of conceptual regions" (2017, p. 4019) That's true, but I put forward the stronger convexity criterion as an empirical hypothesis. This is one aspect where conceptual spaces lead to richer predictions than prototype theory. Another difference is that prototype theory does not build on the geometric structures that are central to conceptual space.

Secondly, in the case when a domain has a Euclidean metric, there are convex tessellations of the domain that are not generated by the Voronoi mechanism, so in this subcase, Thesis V is stronger than the convexity criterion. Thirdly, when a domain has a metric that is not Euclidean, Voronoi tessellations may lead to non-convex regions, so in this case Thesis V and the convexity criterion are in conflict. So if I accept the convexity criterion, I cannot endorse Thesis V at the same time. However, many of Hernández-Conde's arguments assume that I do. When he accuses me of *petitio principio* (2017, p. 4021), he also assumes that I take Thesis V for granted.

Another misrepresentation occurs when he writes: "The convex regions are identified by Gärdenfors precisely with concepts" (2017, p. 4013). This is not correct since already in Gärdenfors (1990), I explicitly state that I only view the convexity criterion as a *necessary* but not sufficient condition for a natural concept.

In sections 5–7, Hernández-Conde argues that if the metric of a space is not Euclidean, then the tessellation into regions generated by a Voronoi classification *could* result in non-convex regions. True, but this is again an empirical question whether concepts that depend on psychological spaces with a non-Euclidean metric actually correspond to non-convex regions.

Hernández-Conde (2017, p. 22) makes a logical mistake when he argues from "convexity is guaranteed under a Euclidean metric" to "the theory requires that the metric underlying our psychological space is Euclidean". This does not follow since convexity can be defined also for other types of metric. There he also writes that "the main argument in favor of a Euclidean metric is that in the case of integral dimensions, a Euclidean metric fits the empirical data better than a city-block metric". This is a misrepresentation, since the fact that a Euclidean metric fits the empirical data better than a city-block metric is used in the psychological literatures as one criterion (among others) for *deciding* whether dimensions are integral or separable (see Gärdenfors 2000, pp. 24–26 for a presentation of some criteria and Johannesson 2002 for an extensive discussion). This criterion is thus a part of a definition and not an argument for the Euclidean metric.

In section 7, Hernández-Conde proposes another distance measure that combines prototypes and the number of examples on which a concept is based. Hernández-Conde's point is to show that with such a distance measure, the resulting regions would not necessarily be convex. This is true, but, again, it is an empirical question which rule for categorization fits the data best. As far as I know, there is no psychological data that supports the categorization method he suggests. My hypothesis is still that convex partitionings have more empirical support.

Another misrepresentation occurs in Hernández-Conde's discussion in section 8.1 of how shapes are represented in conceptual space (in relation to the apple example). I have put forward the prediction that shapes correspond to convex regions of the shape domain. Admittedly, this prediction is not directly testable unless I specify the structure of the shape domain (I discuss some possible approaches in Gärdenfors 2000, section 3.10.2 and Gärdenfors 2014, section 6.3). However Hernández-Conde argues that this prediction is false, since the epicycloid shape of apples is clearly non-convex. This is, however, a clear misunderstanding of how shapes are represented in conceptual spaces. The claim is not that the shape of an object is convex in physical space, but that the *class of shapes* of a category forms a convex region in the shape domain. Thus if an apple A has a particular shape and apple B another shape then any shape *between* that of A and B would also be a example of an apple shape. If apple shapes are (approximations of) epicycloids as generated by the pair equations provided by Hernández-Conde for a range of radiuses r, then this class is trivially convex in the sense that r is the only variable, so that if the shape of A is determined by r_1 and that of B by r_2, then any value between r_1 and r_2 is also generates an epicycloid belonging to the class.

In section 8.1, Hernández-Conde also argues that the concept of a swan is a counterexample to my definition of object categories since swans are either black or white, so the convexity criterion is violated. However, color is not one of the determining properties for 'swan'. Apart from property regions that are part of all categories falling under the superordinate category 'bird', the shape domain is presumably the most characteristic. This means that the representation of the category of swans does not exclude any color. If a yellow bird that in all other respects had the properties of a swan were discovered, it would be categorized as a swan (this is what happened when the black swans in Australia were observed by the European explorers). In brief, this is not a counterexample, but a fairly standard case of object categorization mechanisms.

In conclusion, I am grateful to Hernández-Conde for pointing out so many cases where the convexity criterion *could* be violated. This shows that the criterion is rich in empirical content. The cases that Hernández-Conde discusses can, and should be, empirically tested (see e.g. Jäger 2010). Hernández-Conde does not present any valid counterexamples to the criterion, but only points out they *could* turn out to be false. The upshot is that, in the absence of counterexamples, the features that Hernández-Conde views as problems are instead *strengths* for the theory of conceptual spaces.

Acknowledgements I thank the reviewers for valuable comments on an earlier version of the text. I also gratefully acknowledge support from the Swedish Research Council to the Linneaus environment Cognition, Communication and Learning.

References

Gärdenfors, P. (1990). Induction, conceptual spaces and AI. *Philosophy of Science, 57*, 78–95.

Gärdenfors, P. (2000). *Conceptual spaces: The geometry of thought*. Cambridge, MA: MIT Press.

Gärdenfors, P. (2014). *The geometry of meaning: Semantics based on conceptual spaces*. Cambridge, MA: MIT Press.

Hernández-Conde, J. V. (2017). A case against convexity in conceptual spaces. *Synthese, 194*(10), 4011–4037.

Jäger, G. (2010). Natural color categories are convex sets. *Amsterdam Colloquium 2009, LNAI*, 6042, 11–20.

Johannesson, M. (2002). *Geometric models of similarity* (Lund University Cognitive Studies 90). Lund: Lund University.

Part II
Evolving Concepts

Chapter 6
On the Essentially Dynamic Nature of Concepts: Constant if Incremental Motion in Conceptual Spaces

Joel Parthemore

Abstract Concepts are the means by which we structure our understanding of the world and consequently the primary means by which we encounter it. It is commonly assumed that one of the essential characteristics of concepts – regardless of referent – is their stability, tending toward stasis; and, indeed, it can be hard to see how concepts can otherwise be systematic and productive, in the way they are conventionally taken to be. Even the question has been raised whether concepts *can* change; on some prominent accounts, they cannot. The Unified Conceptual Space Theory (UCST) – an extension of Conceptual Spaces Theory – makes the controversial claim that concepts not only are subject to change over an iterative lifecycle but that, at an underlying level, they are in a state of continuous motion; indeed, they *must be* to function as they do. Mere openness to change is not enough. Even the most seemingly fixed of concepts – mathematical concepts are the paradigm example – can be seen to evolve and continually be evolving as our understanding of mathematics evolves. UCST suggests that concepts possess an intrinsic tension that appears to present a contradiction: to be able to apply in more or less the same way across unboundedly many contexts (systematicity) and to be able to combine coherently with other concepts (productivity), they must be *relatively* stable; and yet, since each new application context is, in some nontrivial way, different from every previous context in ways that do not fit within neat conceptual boundaries, they must adapt each time to fit. In a physical world we have reason to view as ultimately one of fluidity, of processes and motion rather than stable entities, concepts *should* probably have a similar nature.

J. Parthemore (✉)
University of Skövde, Skövde, Sweden
e-mail: joel.parthemore@his.se

M. Kaipainen et al. (eds.), *Conceptual Spaces: Elaborations and Applications*, Synthese Library 405, https://doi.org/10.1007/978-3-030-12800-5_6

6.1 Introduction: How Stable Are Our Concepts, Really?

6.1.1 Concepts and Theories of Concepts

Theories of concepts are attempts, within cognitive science and philosophy of mind, to say what concepts are: i.e., what systematically structured thought is. They seek to lay out the ground rules for the organization of "higher-order" minds capable of stepping back from the present moment to consider it and its contents in light of moments past and moments yet to come. Among the contemporary theories being debated one finds (Fodor 1998) Informational Atomism, (Prinz 2004) Proxytypes Theory, (Gärdenfors 2004) Conceptual Spaces Theory (CST) – and the author's own (Parthemore 2011, 2013) Unified Conceptual Space Theory (UCST), an extension of CST. CST sees conceptual spaces as the analog to physical spaces, with a different space for each conceptual domain, its geometry determined by the *integral dimensions* of that domain[1]; while UCST attempts to show how all the different conceptual spaces described by CST come together in a single, unified "space of spaces" defined by the most basic integral dimensions common to all concepts. UCST comes with a toy software program for generating *mind maps* (Fig. 6.1): visual descriptions of a given conceptual domain (Parthemore 2011, ch. 8).[2] The present paper is largely set within the framework of the latter two theories, though the claims that it makes should resonate well beyond.

For sake of working definition (one that should be acceptable within all the theories mentioned above), let us take concepts to be either the building blocks of systematically and productively structured thought or the abilities by which a certain class of agents – call them *conceptual agents*[3] – are able to engage cognitively with their environment in a systematically and productively structured fashion.[4] Concepts afford them a flexibility of response to that environment akin to that afforded by consciousness.[5] Although various researchers have offered their largely if not strikingly similar lists of the defining properties of concepts (see e.g. Chrisley and Parthemore 2007; Fodor 1998; Laurence and Margolis 1999; Prinz 2004) – which generally if not universally include systematicity and productivity –

[1]... In the case of the color space, they may be taken as *hue, saturation,* and *brightness*.

[2]Proxytypes Theory, Conceptual Spaces Theory, and the Unified Conceptual Space Theory are all derived from Prototype Theory, as developed by Rosch (1975, 1999).

[3]... Or groups of those agents collectively. While some would distinguish between concepts for the individual and concepts for the group or society (see e.g. the discussion in Woodfield 1994), I have argued (Parthemore 2014b) that these represent the same phenomena, structured in essentially the same way, on different levels: i.e., individual, group, society, and possibly species.

[4]This longstanding debate, whether concepts are best understood as (abstract) objects or as abilities – is discussed in the opening pages of Laurence and Margolis (1999) and at some length in Parthemore (2011, Ch. 2).

[5]For a discussion about the links between conceptual agency and consciousness, see Parthemore (2017), which argues that the two are really two sides of one coin.

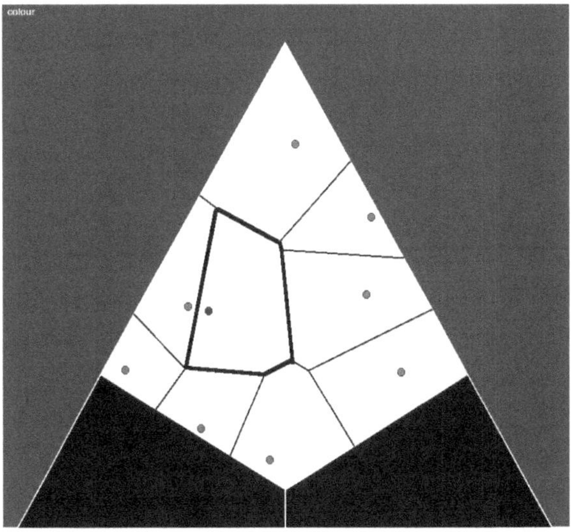

Fig. 6.1 A partitioning of the *space of properties* (one of the three most general spaces within the UCST framework, along with the *space of objects* and the *space of happenings*). The concept "colour" (highlighted) stands in relation to such related properties as size, shape, and density. The person creating the mind map/partition can delve into the "colour" concept to subpartition it, or can add additional concepts to the space surrounding it. Of course, concepts (and so partitions) can also be deleted. (Illustration by the author)

I'm not aware of anyone listing stability or ultrastability among those properties. Nevertheless most researchers would appear to assume that concepts are, at least most of the time, stable[6] entities – and some (notably Fodor 1998) go so far as to argue that concepts do not and cannot change.[7] If most researchers would allow that (at least most) concepts *can* change – within limits[8] – they would also generally hold that, most of the time, concepts do not. Both the resistance to and boundaries

[6]By "stable" I simply mean substantially resistant to change.

[7]This is because, on his Informational Atomism account, concepts are mostly atomic (unstructured) entities defined solely by the things they track in the world. That is to say, Informational Atomism assumes some form of realist metaphysics. One could worry that, unless a realist metaphysics is presupposed a priori, that such an approach avoids the interesting story, which might seem (on an antirealist account, say, such as Immanuel Kant's *transcendental idealism* offers) to be the tangled relationship between concepts and their referents. That is to say, on such accounts, the interaction between concepts and referents is where the action is.

[8]Most researchers would allow that most concepts can change because most allow that beliefs are (partly) constitutive of concepts and not just that (as, so far as I know everyone – including Fodor – allows) concepts are constitutive of beliefs. Because beliefs can and do change – even if they may happen not to – therefore concepts can, and do, change. Note again though the frequently made distinction between individual agent's and societally shared concepts, whereby researchers are often much more willing to acknowledge change in the former than in the latter; see again Woodfield (1994).

of change are important: after all, what use would a concept of "grizzly bear" be if it applied to a type of mammal one day and a kitchen utensil the next? On reflection, it may seem as if this tending-toward-ultra-stable nature is obligatory on concepts, and that nothing more needs to be said.

6.1.2 Conceptual Change

Still – with a possible exception for mathematical concepts like parity and primeness, most researchers would, I think, accept the following:

Thesis 1. Concepts – to function as concepts – must be open to change.

This implies that any concepts that completely cease to be open to change are, metaphorically speaking – except perhaps for historical purposes – dead.[9] That is because those entities which concepts reflect are meant to be, if not changing, then open to change (with the possible exception of *natural-kinds* concepts, which I will come to presently). Mathematical concepts are often thought of as particularly stable; but even concepts like parity conceivably *could* change if e.g. number theory were revised or expanded; after all, at no point can one comfortably announce that one has arrived at the "right" number theory.

The notion of conceptual "death" in turn implies the following, which I likewise take to be non-controversial:

Thesis 2. Concepts may be seen to follow a life cycle of birth, maturation and (at least in certain cases) death.

Corollary. The death of one concept is often the birth of another, or of several others.

Example. When the concept of *phlogiston* was discarded in the late 18th Century, the concept of *oxidation* may be seen to have taken its place. Although the "birth" of the oxidation concept preceded the "death" of the phlogiston concept (which lived on and lives on solely as a matter of historical interest), nevertheless the former may be seen as the natural heir of the latter.

This paper takes the far stronger position – strongly implied by UCST but, so far as I know, not defended elsewhere in print – that concepts are, by their nature, and from a critical and irreducible perspective, in a state of *continuous* (if often only incremental) change. The claim proceeds from what might be observed as a central (albeit seemingly paradoxical) tension in the nature of concepts. On the one hand, concepts – to function as concepts – must be both stable and general ("context free") enough to apply across unboundedly many contexts; systematicity and compositionality would seem to require if not outright presuppose this. On the other, concepts always are applied within specific contexts – each of which is, seemingly unavoidably, different in *some* substantive way from any that have

[9]Cf. the concept of *dead metaphor*, where a metaphor (e.g., "safe haven") is argued not to function any longer as a metaphor.

gone before. That implies that concepts *must* be sensitive to context (i.e., "context sensitive"), adapting to fit each new context as needed: precisely Heraclitus' point when Plato quotes him (*Cratylus,* 402a) as saying "you cannot step into the same river twice". The following seems safe, again, to allow on most accounts:

Thesis 3. Concepts must be just stable enough, but not too stable! Too stable or too dynamic and they cease to function as concepts.

The claims made by all three theses are mainly metaphysical rather than empirical in nature. They provide generally acceptable starting assumptions that, if they are *good* starting assumptions, will be explanatorily productive. If they are poor, they will prove awkward or lead to dead ends.

6.1.3 From Openness to Change to Obligation to Change

These theses together are not, of course, sufficient to require *continuous* change. To get there requires two further ideas: first, that concepts are one thing when we self-reflect on them as concepts – in which case they typically appear as stable representations (often called *mental representations*); and logically quite another when we simply get on with possessing and employing them non-reflectively as, seemingly, most of the time we must do – in which case they might seem to be something else, something non-representational[10] and action-based (for we are using them, not reflecting on them).[11] Actions are, by nature, things in motion; and motion implies (if not requires) change. These two contrasting perspectives do a great deal, I think, to explain and resolve the debate over whether concepts

[10]Representations seem, to many enactive researchers among others, to imply if not outright require an agent to recognize them *as* representations: i.e., to take a *representational stance* toward them. People often use the term "representation" loosely, without ever bothering to define it. My longstanding inspiration on how *I* wish to use the term comes from Inman Harvey: "The gun I reach for whenever I hear the word *representation* has this engraved on it: ′When *P* is used by *Q* to represent *R* to *S*, *who is Q and who is S*?′. If others have different criteria for what constitutes a representation, it is incumbent on them to make this explicit" (Harvey 1992).

[11]For an extended discussion, see Parthemore (2011, pp. 110–112) and Parthemore (2013). Note the paradox here: the moment that we reflect on what it is like to possess and employ concepts non-reflectively, we bring them into the realm of self-reflection; in the very action of trying to catch them "in the act", we pull them *out* of that act and into the abstract realm of reflection. Reflecting on what is going on when we are not reflecting on what is going on threatens – like Russell's paradox – an unresolvable oscillation between two opposing views. As Zoltan Torey writes: ". . .we may take an object and just by focusing on it we notice almost at once that it (the content component) begins to recede and become overlaid by the nonthematic sensation that the whole experience is our own doing. However, this same sense of self-contribution, too, begins at once to fade, allowing the attention to swing back once more to the object in focus, from there to fade in turn, accentuating the self-sensation once more before the attentional pendulum swings back to the object again" (Torey 2009, p. 112). Torey thinks this oscillation can be resolved by careful logical analysis and the underlying truth arrived at; I do not.

are "really" representations or abilities. In truth, both perspectives are needed, and neither can be reduced to or otherwise reconciled with the other. If one allows these two contrasting perspectives, then one will further allow that the apparent stability of concepts, when we reflect on them, need not and almost certainly does not reveal their full nature.

The second requisite idea, which I take to be largely uncontroversial (at least until considered in its full implications!) is that concepts are massively interconnected and – with care to avoid too close of a dictionary metaphor[12] – interdefining. Not every agrees with such a view: Fodor, pointedly, views concepts (which he understands as atomic symbols) as strictly independent of one another, whereby "it's plausible *prima facie* that 'a' might refer to *a* even if there are *no* other symbols" (Fodor 1998, p. 54).[13] The idea is that knowing about weddings or funerals may require, *inter alia,* knowing about flowers, which may require understanding what a rose is, which may require recognizing red, which may also be connected to one's understanding of fire hydrants, which may be connected with one's concept of dogs (at least in the popular American imagination), which requires understanding what a pet is, which is why – even if one has never heard or used the expression before – one knows immediately what Fodor means when he uses his favourite conceptual example of a *pet fish.*[14] Fish is a type of seafood, which relates to sushi, which – if one has had a bad experience – may connect with one's concept of food poisoning. At no point does one reach the point where one may safely stop; the regress goes on as long as patience and cognitive resources allow.[15] Think of concepts like a spider's web: pull on any one strand of the web, and the entire web vibrates, even if no one is there to observe it (or to observe it all). Alternatively, consider each new conceptually mediated experience like a pebble in a body of water, sending

[12]Classical definitionist accounts are massively out of fashion, and for good reason, as both Fodor (1998) and Jesse Prinz (2004) are at pains to point out.

[13]Note, too, that, for Fodor, concepts are *lexical* concepts (or compounds of lexical concepts), whereas I have consistently taken the view that concepts and language pull apart (another reason to avoid anything even smacking of definitionism): see e.g. Parthemore (2014b). Consequently I think – *pace* Fodor, but in keeping with the so-called *animal concepts philosophers* (see e.g. Allen 1999; Newen and Bartels 2007) – that pre-linguistic infants past a certain point in development along with some number of non-human species possess conceptual agency.

[14]Fodor takes the pet fish as proof that concepts are not prototypes, on the argument that prototypes are compositional while concepts are not: for a pet fish is neither a prototypical pet nor a prototypical fish, never mind the intersection of the two. I prefer to take a different lesson: that concepts compose in a different way and along different paths from language, and that one does not arrive at the concept of pet fish simply by combining the lexical concept of pet with the lexical concept of fish. In short: an overly linguistic view on concepts seriously distorts one's view of them.

[15]Some years ago, I wrote a short-story writing environment for my master's thesis. The program invited the user to create a story world and populate it with characters, actions, objects, etc. A colleague, trying out the program, quickly found himself 30 levels down in recursion, describing what toilet paper is, to explain what a toilet is, to describe a spaceship, to explain something that described something that contained something else – all to create his first character, so he could write his first sentence.

ripples to the farthest reaches – even if only some of those ripples are visible to the naked eye. *If concepts and conceptual frameworks are massively connected in this way, then a change anywhere in the system will produce at least marginal movement throughout the system.* To avoid this, one must argue either that concepts in general or some concepts or clusters of concepts in particular[16] are more weakly connected: that is, that they are substantially context free, that they do *not* exist within a unified "space of spaces" such as UCST suggests.

The central claim of this paper, to be developed in the pages that follow, is that concepts – if they are to function as concepts at all – are, at least when we are not looking at them, moving targets. They are in a state of constant if incremental motion where each application of a concept (and we apply concepts constantly through our waking lives, along with no small part of our sleeping ones) causes ripples throughout the system. The claim further is that this should be true on every level on which one may talk about concepts – individual, group, society, even species – albeit on different time scales (Parthemore 2014b). Even the most seemingly fixed of concepts – say, mathematical concepts of primeness or parity – may be seen to evolve *and continuously be evolving,* for the individual and for society. Failure to be aware[17] of change should not be taken as lack of change – not if the circumstantial evidence and logical arguments in favour of (continuous) change are sufficiently strong, as I will try to convince the reader they are.

6.1.4 What I Am Not Saying

I am *not* claiming, as Heraclitus is commonly read, that "all is change and nothing stable" – which would reduce conceptual stability to at best a comforting illusion. This would be as wrong, I think, as to claim – *per* Parmenides in his famous poem – that nothing ever changes because change is illusion. Plato's solution was to adopt Parmenides' perspective for the world of his perfect Forms while allowing Heraclitus' perspective for the world we observe, where the Forms appear as but shadows on the cave wall. My own solution is to reject Plato's Forms in favour of Kant's noumena and to say that change and stability can co-exist simultaneously, in the same moment in the same entities in the same world, depending on whether one is stopping to reflect on the world or trying to catch that world simply in the act of being. There simply is no one fact of the matter.[18]

---

[16]Natural kinds concepts, if they exist, would be the obvious example. Needless to say, the claims made here argue against them; see Sect. 6.3 below.

[17]...Which is to say, reflectively self-aware – if, in keeping with the phenomenologists, one distinguishes between "bare" conscious awareness (pre-reflective) and conscious awareness of conscious awareness (self-reflective); see e.g. Gallagher (2007).

[18]I wish to thank Sylvia Wenmackers (email) for reminding me of Heraclitus and Parmenides and Plato's response to them.

Neither am I – by arguing that concepts are constantly moving – denying the possibility of discontinuities: i.e., sudden, abrupt changes. Clearly, not all conceptual change is gradual, either for individuals (remember the moment you realized that nothing *was* hiding in the closet?) or societies or civilations (consider the conceptual changes set in motion by the discovery of microorganisms and pathogens). Particularly at the societal level, conceptual frameworks tend to resist change (beyond the necessary) for as long as they can, building up pressure at the boundary regions between one concept and another or one conceptual domain and another until a critical threshold is reached – at which point entire conceptual spaces may give way to make room for something new.[19] A sudden insight may set a whole cascade of changes in motion.

Finally and perhaps most importantly, I am *not* claiming that the conceptual agent is, or even can be, aware of all the conceptual change – either within herself or in her society. The spider is aware of the slightest vibration in the furthest corner of the web, but the human observer loses track of the ripples in the water long before they die away. We should be particularly wary of what we conclude "must" logically be the case when we attempt to set our individual and collective (self-)reflections aside – if such a "perspective without a perspective" is even coherent! – and describe the world or the entities in it "as they really are". The claims put forth in this paper are not meant to be a priori; rather, they are meant to hang together in a coherentist-like fashion, to facilitate explanations that might otherwise seem difficult or contrived. They provide no formal proof but only accumulate circumstantial evidence in their favour. A theory of concepts that allows concepts to be in continuous motion is, I think, a simpler, more elegant, more straightforward account than one that does not – as, again, I hope to convince the reader.

6.1.5 The Remainder of This Chapter

Section Two presents what I call the Illusion of Non-motion and argues via analogy that appearance of non-motion cannot be taken at face value. Sections Three and Four present the corresponding Argument from Motion and consider the role of abstraction in altering our view of that motion. Section Five discusses the theoretical and practical consequences of embracing the views expressed in this chapter. Section Six reviews the discussion and offers parting thoughts.

[19]For a visualization of this – with thanks again to Sylvia Wenmackers – see https://vimeo.com/231498722, accessed 2018-02-27.

6.2 The Illusion of Non-motion

Appearances can be deceiving. Just as apparent motion can be an illusion (Fig. 6.2); so, too, can apparent non-motion.

Plants are generally understood by lay persons – and often enough described by researchers – as non-moving, as opposed to animals (and microorganisms), which move around. But as any number of researchers have pointed out, the main difference in movement between plants and animals (beyond the existence, in most plants, of a root system, which fixes location and severely limits range of movement) is one of time scale. Plants *appear* not to move because they move, most of the time, on a much slower time scale than we do. So when we observe them naively, they appear motionless; when we change viewing perspective – most conveniently, by using time-lapse photography – they can be seen to move around quite actively, even purposefully, orienting themselves toward light (e.g., sunflowers) or water (the root system of many plants), or away from certain environmental irritants (e.g., *Mimosa pudica*); or capturing food, in the case of carniverous plants (e.g., Venus fly trap).[20]

Consider the first time, as a child, that you looked at a sample of pond water under a microscope. Again, a change of perspective takes something that at first *appears* inert and shows it to be full of motion. Or consider – likely with memories that evoke childhood – looking out the window of a fast-moving train at nearby vegetation. Intellectually you realize that the vegetation is rushing past you (or you

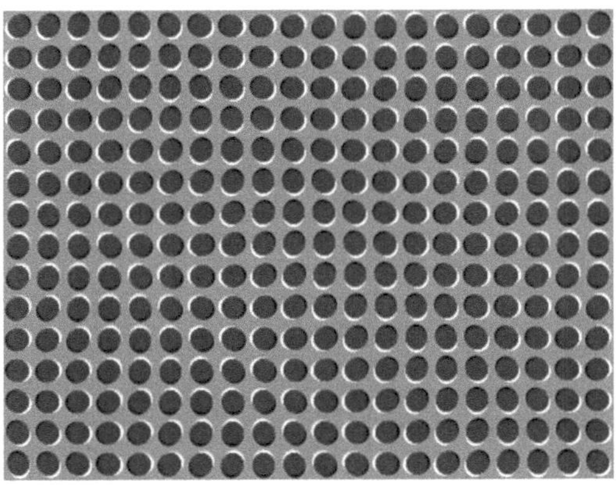

Fig. 6.2 An illusion of motion: for most persons, the picture appears, when examined closely, to be waving back and forth like a flag in a slow breeze. Just as something can appear to be moving, even when rational consideration says that it's not (and cannot be); so, too, something can appear *not* to be moving even when rational consideration says that it is (and must be). Image reproduced from Wikimedia Commons: http://commons.wikimedia.org under the Creative Commons license

[20] A YouTube search on "plant movement" and "time-lapse" yields a number of striking videos.

past it) the same way as objects further in the distance; but because of its closeness, the "moving" vegetation (which is "moving" far more quickly than the background!) resolves into a stable, abstract pattern: a "fixed" tapestry of colour.

Finally, consider that, just as one can create a simple animation – an illusion of motion – by drawing a figure on a series of sheets of paper and then riffling through the sheets quickly, so that the figure may be seen to run, or skip, or jump, or play ball, and so on[21]; so, too, one can achieve a complementary effect by drawing a figure on a series of sheets of paper, where *the entire figure is not to be found on any one of the sheets*: that is, various sheets fill in details that others lack, or add details that are seen to disappear. By riffling through the sheets quickly, a stable image is formed that is not to be found on any one or another of the sheets. The figure can even shift position somewhat between sheets ("move around"), and yet the resulting image appears stable and motionless.

The seeming motionlessness of concepts arises not because, like the sunflower, they are moving too slow; nor because, like the vegetation viewed from the train, they are moving too fast. Neither is it because, like the pond water, we are failing to look closely enough (in that case, by using sufficient magnification) – because, no matter how intently we self-reflect on our concepts, whatever they happen to be concepts of (concrete or abstract objects, concrete or abstract actions or happenings,[22] or properties of these things), stable representations are what we seem to find. Concepts are more like the partial figure on the sheets of paper, for concepts live in each instance of their application, even if what we recognize most readily as a concept is the seemingly stable image that emerges from consideration of all the instances of application, collectively: that is, that which abstracts away from all the instances of application. Concepts resemble the toys from fairy tales[23] that only move (or only can move) when no one is watching; one recognizes their movement from the traces they leave behind. One looks, looks away, looks back again, and something has changed.

What is important to remember is that, if this chapter is correct, there is no one single fact of the matter about whether concepts are in continuous motion[24]; as the preceding discussion attempts to make clear, what counts as motion in the first place is often dependent on *perspective*: an irreducible and ineliminable characteristic of conscious, conceptual agents, who cannot help but take a perspective. In his "Three-

[21]This is often an early exercise in teaching animation.

[22]Note, however, that there *is* an important difference between concrete and abstract entities when it comes to motion and change, which I will discuss in Sect. 6.3 below.

[23]...Or the Weeping Angels from the British science-fiction series DR. WHO: statues of stone that literally come to life in the blink of an eye.

[24]This goes back to my point about the ongoing debate over whether concepts are (or are best understood as) representations or abilities, with each side assuming that one side or the other *must* be right. Rather, which perspective is primary and which secondary depends on circumstances, on the needs of the moment.

part invention", Douglas Hofstadter (2000, p. 30) relates the following Zen koan[25] through his character Achilles:

> Two monks were arguing about a flag. One said, "the flag is moving." The other said, "the wind is moving." The sixth patriarch, Zeno, happened to be passing by. He told them, "not the wind, not the flag; mind is moving."

6.3 A World in Motion

The world around us as a whole – the world in which we find ourselves embedded and constantly (on some level) interacting (whether we are aware of doing so or not), the world from which we derive most if not all of our concepts, often strikes us in many ways as itself unchanging and unmoving. That, too – as many philosophers from Heraclitus onward, and many empirical researchers as well have claimed – is an illusion. The world, they say, is one of continuous motion and change, and it is only observation and reflection that appear to fix it in place: think of quantum mechanics (with its ineliminable role for the observer) and Schroedinger's cat.

Those things that seem most resistant to change *do* change, in time. On a geological time scale, lakes and entire river systems appear and disappear in the geological blink of an eye. Mountain chains like the Appalachian Mountains on the East Coast of North America are built up, worn down, built up again, and worn down again. Continents break or drift apart,[26] or collide together.

But surely the Earth beneath my feet is fixed enough? ...It is, relative to my position on it. At the same time both of us are rotating around the Earth's axis at a speed (given my present latitude) of about 958 km/hr[27] while moving around the sun at a speed of about 107,279 km/hr. The solar system orbits the centre of the Milky Way galaxy at about 828,000 km/hr; once around takes close to a quarter of a billion years.[28] The Milky Way galaxy itself, of course, is moving; on present course, it is set to collide with the Andromeda galaxy in four billion years.

Concepts – on just about any account and regardless of one's metaphysical starting point – are meant to reflect the world as accurately as possible. After all,

---

[25] A *koan* is a kind of mini-story used in Zen Buddhism to break people out of their conventional thinking habits and encourage spiritual awakening.

[26] ...Which is why traces of the earlier Appalachian range are found on three continents.

[27] I would need to be standing at one of the poles for my rotational speed to be approximating to (though never actually reaching) zero.

[28] These speeds are relative to our viewing perspective, of course. One of the striking consequences of Relativity Theory is that for every object in the universe, depending on one's vantage point, that object may be seen to be moving any speed at all from zero (rest) up till arbitrarily close to (though never reaching!) the speed of light: i.e., it's the *relative* speed that determines the observed speed; there is no simple fact of the matter what velocity something is traveling because there is no one, privileged perspective. From our vantage point on Earth, the most distant observable galaxies are moving away from us at speeds approaching the speed of light.

they are the means by which we structure our understanding of the world; they are the instrument through which we approach and engage with it. If Heraclitus is right and the world is one of continuous motion and change then concepts, to reflect that world accurately, should be in continuous motion and change as well – otherwise, the presumed isomorphic relations between the two will quickly break down. (How well can the whole match if the parts do not?)

Unless one presupposes the existence of *natural kinds*[29] – (unchanging and unchangeable) concepts that "carve nature at its joints" (to borrow Plato's famous phrase), picking out fixed regularities in the world,[30,31] then one will at least be open to the idea that concepts impose boundaries onto what are, from a conceptual perspective, underlying (fluid) continuities. That is to say, although the world severely constrains the categories we impose on it, it need not (and arguably should not) be seen as providing those categories ready made (Parthemore 2011, pp. 58–59). Such a view gains credance from the widely reported phenomenon (in the psychology literature) of *categorical perception*, whereby a quantitatively evenly spaced set of stimuli (such as tones on a scale) are nevertheless perceived as falling into precise categories on one side or another of a binary divide (*x*, not *x*): i.e., "stimuli related to a specific category are perceived as indistinguishable, whereas stimuli from a 'nearby' category are perceived to be entirely different.... In color perception, for example, different shades of green are perceived to be more similar than green and yellow even though the wavelength differences are no larger" (Gärdenfors and Williams 2001, p. 387; see also Harnad 1990a).[32]

Take away natural kinds and we, as conceptual agents, become fluid motion in a world of fluid motion with no set boundaries – understanding both our world and ourselves as stable, well-bounded entities to the extent it suits our needs. We think of ourselves as the same person we were in our youth, but our ideas, beliefs, understandings change – most importantly, sometimes, our understanding of ourselves and what we see as our place in the world (for we are each, in an important way, the center of our *own* worlds). We remember those things that give

[29]... As does e.g. Brian Ellis (2005).

[30]The elements in the periodic table are frequently offered as examples.

[31]For all that natural kinds philosophers look to him for support, W.V. Quine may be read as putting the concept up to question: e.g. (1969, p. 13), "it is reasonable that our quality space should match our neighbor's, we being birds of a feather; and so the general trustworthiness of induction in the ostensive learning of words was a put-up job. To trust induction as a way of access to the truths of nature, on the other hand, is to suppose, more nearly, that our quality space matches that of the cosmos. The brute irrationality of our sense of similarity, its irrelevance to anything in logic and mathematics, offers little reason to expect that this sense is somehow in tune with the world – a world which, unlike language, we never made." Quine offers the hope that, in a "mature" science, the concept of *kinds* would disappear, and only the *natural* remain.

[32]There is, of course, no a priori reason to presuppose that categorical perception generalizes to *all* concepts; nevertheless, the widespread reporting of the phenomenon in different sensory modalities, along with the common argument from the empiricist tradition that concepts are *grounded* in perception (Harnad 1990b) (or, more accurately perhaps, in sensorimotor engagements: see e.g. Gallese and Lakoff 2005; Parthemore 2014c), raises that as a strong possibility.

us a sense of continuity and forget or disregard those things that seem quite alien to our present-day selves until something brings them back to mind – when suddenly we remember the child terrified of the shadows in the closet, maybe even are that child for a moment again, as Ray Bradbury (2014) suggests in his short story, The thing at the top of the stairs. Of course this tension between being the same/not being the same is precisely what drove the ancient Greek debate and presented, for some, a paradox.

In discretizing the world, concepts impose a *degree* of stability onto its continuous motion, while never eliminating it entirely. They reify processes into objects, at the same time as abstracting away from their fluid nature, simplifying and giving them a more concrete appearance, attempting to fix them in place, taking "snapshots" that constantly need to be "touched up" (because something in the "picture" has moved). If the reasoning is anything like correct, it raises the provocative question: how much is our view of a relatively stable and predictable world a conceptual imposition?

6.4 Concepts and Conceptual Spaces in Motion

So the conclusion should be that concepts are always in motion, jostling each other around, but not *as much* motion as the world they reflect: that is, they slow the motion down without ever stopping it. This follows naturally if one views concepts holistically, as part of a highly intertangled conceptual framework that is constantly being updated – as I have suggested one should do; rather than (as *per* Fodor's account) independent atoms bearing no relation to each other but only to the "external" world. All one's experiences to date – along with all the experiences one reads or hears about from others (i.e., experiences second hand) – are brought to bear (in whatever often highly indirect way) on the concepts one is applying in the present context and the reshaping of those concepts in their application: which is to say that all one's conceptually mediated experiences to date perturb one's present perspective, and conceptual agency is all about perspective (viewing the world one conceptually mediated way, and not another, where no one perspective is privileged above all others).[33]

At the same time that concepts discretize the world, imposing a degree of stability onto its continuous motion, *they do the same thing to themselves.* That is to say that our concepts, as and when we reflect on them – when we form concepts *of* our concepts – are themselves reifications, simplifying themselves, giving themselves

[33]This is not, of course, to deny that many perspectives are not useful or simply (for practical purposes) *wrong*, because they fail to reflect the world and one's interactions with it suitably. One can believe the world is flat, but that mean that one can treat it as such under all circumstances, without unfortunate consequences following! (This is not is not to open any discussion into the ontological status of truth, which is – it goes without saying – beyond the scope of this paper.) My thanks to Blay Whitby for private discussions on this point.

Fig. 6.3 The author in a hall of mirrors. (Self-portrait)

a more concrete appearance, attempting to fix themselves in place. Again, it seems logical that they should be one thing when we are simply getting on with using them, quite another when we stop to look at them: cf. Michel Polanyi's (1958) distinction between *focal* and *subsidiary* awareness, itself evocative of a much earlier discussion from Heidegger on hammers (1978, p. 98).

Just as we can, and do, form concepts of our concepts: our concept of our concept of a dog ("how would I describe a dog?"), our concept of our concept of democracy ("what is democracy?"), our concept of a concept itself ("what does it mean to have a notion of whatever?"); so, too, we can form concepts of our concepts of our concepts.[34] Quickly, however, we need to stop. Why? ... Because the concepts become too abstract, too far removed from their origins: an endlessly receding series of reflections, each level distorting the "image" a little more so that the point of departure might get lost entirely (Fig. 6.3).

The more abstract ("higher order", concepts of concepts of...), the more *relatively* stable the concepts become[35] at the gain of generalizability but the loss,

[34]Nothing about direction of flow – bottom-up vs. top-down – should be implied. Indeed, it is one of the standing claims of UCST that nearly all mature concept formation is a product of both bottom-up (unconsciously/preconsciously association-driven) and top-down (reflectively/intellectually driven) processes.

[35]... Which is not to say that they cannot change – even (on occasion) change quickly! When they *do* change though it tends to represent a more radical departure than changes in our more concrete concepts: the analogue to (Kuhn 1970, 1990) talk of *paradigm shifts*.

at least in most cases,[36] of some corresponding degree of applicability to the world. Meanwhile the less abstract ("lower order", concepts of non-concepts), the more *dynamic* the concepts become, the less generalizable but the closer (seemingly) to the physical world.[37]

Of course a theory of concepts – such as Conceptual Spaces Theory, the Unified Conceptual Space Theory, Proxytypes Theory, or Informational Atomism – is itself a higher-order conceptual entity well along that continuum of abstraction and so should not be taken to be too true to the "real" nature of concepts, whatever that may be (besides, to some substantive degree, dynamic). Instead, one should look for particular theories of concepts to be useful in some contexts and not in others, applicable to some questions but not others[38]: never truly universal.[39] The moral of the story is that one should choose the conceptual tool – or theoretical framework – most appropriate to the task at hand as judged by apparent adherence to the reality of that task and explanatory productivity.

6.5 Consequences Theoretical and Practical

If concepts *are* in a state of continuous if incremental motion, what is the conceptual fallout? First, theorists working with models of conceptual frameworks (or, as the area is often if misleadingly referred to in cognitive science, *knowledge representation*[40]) need to take account of time in a much more active way: building a time dimension into the model; building models that change over time and *move,* with some apparently degree of autonomy; building conceptual spaces that change and adapt; thinking less in terms of, say, stable mind maps and more in terms of GPS navigation devices; the map created by these devices is updated on the fly: if the driver misses a turn, the directions immediately and silently change as though nothing had happened. If one is building models of moral agency and the moral frameworks possessed by purported moral agents – such as "intelligent" autonomous artefacts, as I discuss in Parthemore and Whitby (2012, 2013, 2014) (see also e.g. Nyholm 2017) – then one needs to treat that moral agency and those moral frameworks as moving targets. Meanwhile, the very theories of concepts or knowledge that they are working from need to be not just open to change but

[36]One's concept of what a concept itself is – or, if you will, a "notion" or an "idea" or a "thought" – is a necessary place holder but little more, until something more concrete is attached to it.

[37]Compare one's concept of an animal with one's concept of a bear with one's concept of a particular bear: say, the bear that chased you down the trail last summer.

[38]For example, much of the application of Conceptual Spaces Theory to date has been in the field of robotics; see e.g. Chella et al. (2004, 2008).

[39]...Though some, again, may turn out not to be so useful or simply wrong.

[40]Whether all forms of conceptual understanding can usefully be described as "knowledge" is up for debate; but I hope already to have given readers reason to question the unqualified description of conceptual understandings as "representational".

actively seen as evolving entities: works in process. Often, both in publications and particularly in grant applications, one finds a great deal of emphasis on the goal, with what might be seen as exaggerated claims to providing the final, "correct" answer. As John Stewart so eloquently describes in (1995), there is no obvious reason to expect to find one final, "correct" answer for most phenomena of sufficient complexity.

The consequences are not only abstract and theoretical. Researchers need to "do" AI and to build AI systems in a radically new way. Conceptual Spaces Theory is already part of a broader movement, well represented in the enactive community, toward rejecting both *good old-fashioned AI* (GOFAI), with its strongly rationalist bent and its focus on top-down-driven symbolic approaches (including so-called *physical symbol systems:* see e.g. Newell 1980); and connectionism, with its strongly bottom-up association-based approach (including "pure" artificial neural networks with no local but only distributed representations, and the like). Rather than arguing over whether *anything* needs to be built ("hard wired") into the system to begin with – whether it can start with a truly "blank slate"[41] – the question becomes, what is the minimum amount that must be built in to achieve maximum flexibility of response to the environment, maximum capacity for learning/adapting, and minimum overfitting? With regard to Conceptual Spaces Theory in particular, robots need to be built less with fully fledged conceptual spaces built in (see e.g. Chella et al. 2008) and more with *minimally partitioned* spaces (Parthemore 2014a, p. 173)[42] that the robot then develops: that is to say, everything the robot does should perturb, and so force an update of, its conceptual spaces in their entirety; its unified space should be in a state of constant flux.

If self-driving cars are truly to be our collaborative partners rather than fully autonomous agents (Nyholm and Smids 2017) – which might be undesirable on both practical grounds (it's unclear whether artifacts *can* be autonomous in the strongest sense; many, including Jordan Zlatev (2001; 2002), have given reasons to think that might *not* be possible[43]) and ethical ones (what if the vehicle comes to the "wrong" conclusion, from the passengers' point of view?) – then one should build them to update themselves and their conceptually structured understanding[44]

[41]Arguably one of the key problems with GOFAI systems is not that knowledge is built in from the start but just how *much* gets built in: i.e., so much as to constrain severely its capacity to develop "cognitively".

[42]The core argument from that paper is that conceptual agents are born, not with any innate concepts, but with certain innate *protoconcepts*: in particular, protoconcepts for "object", "happening", and "property" that partition the subsequent unified space and force more or less all concepts into one of these broad proto-categories. We cannot make sense of the world and ourselves within that world *except* in terms of (concrete and abstract) objects, (concrete and abstract) happenings, and properties of both; it truly is inconceivable to imagine dividing the world along different lines.

[43]Clearly, treating an agent as fully autonomous when it's not risks dangerous consequences, as described in Parthemore and Whitby (2014).

[44]Perhaps, the reader suggests, their understanding is structured in some other way than concepts – perhaps, but then the burden is on the reader to explain how, in a way that falls outside the dominion of concepts as they are defined here. Otherwise, concepts are simply going by a different name.

of their environment continuously – and sensibly! – in interaction with their driving partners and environment (in, needless to say, as flexible but transparent a way as possible: e.g., through being open to a suitable degree of interrogation). The same applies to robots and other forms of automation in the workplace, at least if one wishes the remaining human workers to view the systems as partners rather than competitors (Danaher 2017) and reduce the risk of costly lawsuits when something goes wrong because the situation was not anticipated by the system designers who hard-wired everything in. That is, the system's capacity to learn must be both robust and, in important if hard to specify ways, human-like; and that requires, at minimum, having a mind (or "mind") in motion.

Finally, are the claims made in this chapter empirically testable? Yes and no. They are, in the main, metaphysical in nature: by which I mean, they concern our *axioms* or starting assumptions. Like other metaphysical claims, they are only open to indirect empirical testing: an accumulation of circumstantial evidence, if you will, as opposed to any kind of proof. In Parthemore (2015) I lay out a detailed plan for empirical testing of CST and UCST, on the argument that, if CST/UCST-based mind-mapping software benchmarks better than traditional mind-mapping software on such key factors as

- readability of maps between subjects,
- continued readability of maps over time (between subjects, and with the same subject),
- clarity of maps to experts in the domain,[45]
- speed and apparent accuracy of task completion,
- ease of using and declared comfort with the provided tools[46] (as judged by post-experimental questionnaires/interviews), etc.

...Then one has good circumstantial evidence in favour of CST/UCST. Contrarily, if it benchmarks more poorly then parts of the underlying theory are either missing or wrong, or the cognitive theory behind traditional mind-mapping software really is superior. Beyond that, the question, ultimately, is whether one can explain empirical observations better – and, say, build better robots! – if one assumes that concepts change, and keep changing, and *must* keep changing, in the ways I describe.

Whatever form that structuring takes, it will presumably still need to be systematic and productive; otherwise, it will fail to generalize appropriately.

[45]All three of these are notorious problems with traditional mind maps, for all of the miraculous claims made on them (see e.g. Sharples 1999, also Parthemore 2011, ch. 8).

[46]...Especially given the additional restrictions imposed by the software; one of the standard complaints against traditional mind-mapping software is that it's massively underconstrained: users can throw down links anywhere and connect them with nearly any sort of arcs.

6.6 Conclusions

With the exception of Fodor, most philosophers, psychologists, and cognitive scientists allow that concepts *can* change. The majority would probably allow that concepts, to function as concepts, must be open to change – certainly if the world they are meant to reflect changes, which it seems that it often if not indeed continuously does. This chapter makes the far stronger and controversial claim that, when we are not looking at our concepts but simply applying them – that is, when considering them more practically and concretely, rather than abstractly – they should be understood as in a state of constant motion themselves. The apparent non-motion of concepts is – taken at face value – at best an incomplete story. What concepts do is to simplify and abstract away from the world at the same time they assist the conceptual agent to be flexible in responding to its environment. They do the same thing to themselves, whether one is talking concepts of concepts or theories of concepts or simply thoughts about thoughts: fixing themselves in place, giving themselves a more concrete, most stable appearance. Viewing concepts in this way, one is inclined not just to build more dynamic models of higher-level cognition but to view static snapshots of such cognition in a new light: as "images" that, if we riffle through them quickly, we can start to perceive the motion that we cannot – due to limitations of perspective – view directly. The individual snapshots are not the conceptual spaces just as, on reflection, the particular immediate application of a concept is not the concept. The movie, not the stills: that is where our conceptual spaces best may be seen as residing.

What I have not discussed, of course, is all the forms that conceptual change can take. Can a taxonomy of forms of change be provided; and, if it can, will it prove useful? How much depends on the weights on various quality dimensions, and how much are these weights assigned (as suggested by CST) or emergent (as UCST implies)? Just what happens when dimensions are added or removed, or when entire conceptual substructures collapse? In any case, conceptual spaces theory, in all its forms, offers rich ways for analyzing that change.[47]

References

Allen, C. (1999). Animal concepts revisited: The use of self-monitoring as an empirical approach. *Erkenntnis, 51*(1), 33–40.

Bradbury, R. (2014). *The Toynbee Convector*. London: Harper Collins Publishers.

Chella, A., Coradeschi, S., Frixione, M., & Saffioti, A. (2004). Perceptual anchoring via conceptual spaces. In *Proceedings of the AAAI-04 Workshop on Anchoring Symbols to Sensor Data* (pp. 40–45). Menlo Park, CA: AAAI Press.

Chella, A., Frixione, M., & Gaglio, S. (2008). A cognitive architecture for robot self-consciousness. *Artificial Intelligence in Medicine, 44*(2), 147–154.

[47] My thanks to an anonymous reviewer for this point.

Chrisley, R., & Parthemore, J. (2007). Synthetic phenomenology: Exploiting embodiment to specify the non-conceptual content of visual experience. *Journal of Consciousness Studies, 14*(7), 44–58.
Danaher, J. (2017). Will life be worth living in a world without work? *Science and Engineering Ethics, 23*(1), 41–64.
Ellis, B. (2005). Physical realism. *Ratio, 18*(4), 371–384.
Fodor, J. A. (1998). *Concepts: Where cognitive science went wrong*. Oxford: Clarendon Press.
Gallagher, S. (2007). Phenomenological approaches to consciousness. In M. Velmans & S. Schneider (Eds.), *The Blackwell companion to consciousness* (pp. 686–696). Chichester: Wiley.
Gallese, V., & Lakoff, G. (2005). The brain's concepts: The role of the sensory-motor system in conceptual knowledge. *Cognitive Neuropsychology, 22*(3–4), 455–479.
Gärdenfors, P. (2004). *Conceptual Spaces: The Geometry of Thought*. Cambridge: Bradford Books. First published 2000.
Gärdenfors, P., & Williams, M.-A. (2001). Reasoning about categories in conceptual spaces. In *Proceedings of the Fourteenth International Joint Conference of Artificial Intelligence* (pp. 385–392). Morgan Kaufmann.
Harnad, S. (1990a). Introduction: Psychophysical and cognitive aspects of cognitive perception: A critical overview. In S. Harnad (Ed.), *Categorical perception: The groundwork of cognition* (pp. 1–25). Cambridge/New York: Cambridge University Press. First publication 1987.
Harnad, S. (1990b). The symbol grounding problem. *Physica D: Nonlinear Phenomena, 42*(3), 335–346.
Harvey, I. (1992). Untimed and misrepresented: Connectionism and the computer metaphor (CSRP 245). http://www.sussex.ac.uk/Users/inmanh/harvey92untimed.pdf. University of Sussex (UK) Cognitive Science Research Papers (CSRP) series.
Heidegger, M. (1978). *Being and time*. Oxford: Wiley-Blackwell. First published in 1927 as "Sein und Zeit".
Hofstadter, D. (2000). *Gödel, escher, bach: An eternal golden braid*. London: Penguin. Twentieth anniversary edition.
Kuhn, T. (1970). *The structure of scientific revolutions*. Chicago: University of Chicago Press.
Kuhn, T. (1990). The road since Structure. In *PSA: Proceedings of the Biennial Meeting of the Philosophy of Science Association* (Vol. 2, pp. 3–13).
Laurence, S., & Margolis, E. (1999). Concepts and cognitive science. In E. Margolis & S. Laurence (Eds.), *Concepts: Core Readings* chap. 1, (pp. 3–81). Cambridge, MA: MIT Press.
Newell, A. (1980). Physical symbol systems. *Cognitive Science, 4*(2), 135–183.
Newen, A., & Bartels, A. (2007). Animal minds and the possession of concepts. *Philosophical Psychology, 20*(3), 283–308.
Nyholm, S. (2017). Attributing agency to automated systems: Reflections on human-robot collaborations and responsibility-loci. *Science and Engineering Ethics, 24*(4), 1201–1219.
Nyholm, S., & Smids, J. (2017). Automated cars meet human drivers: Responsible human-robot coordination and the ethics of mixed traffic. Under review.
Parthemore, J. (2011). *Concepts enacted: Confronting the obstacles and paradoxes inherent in pursuing a scientific understanding of the building blocks of human thought*. Ph.D. thesis, University of Sussex, Falmer, Brighton. Available from http://www.sussex.ac.uk/Users/jep25/papers/thesis/thesis-a4.pdf.
Parthemore, J. (2013). The unified conceptual space theory: An enactive theory of concepts. *Adaptive Behavior, 21*, 168–177.
Parthemore, J. (2014a). The case for protoconcepts: Why concepts, language, and protolanguage all need protoconcepts. *Theoria et Historia Scientiarum, 11*, 159–178.
Parthemore, J. (2014b). Conceptual change and development on multiple time scales: From incremental evolution to origins. *Sign System Studies, 42*, 193–218.
Parthemore, J. (2014c). From a sensorimotor to a sensorimotor++ account of embodied conceptual cognition. In *Contemporary sensorimotor theory* (Studies in applied philosophy, epistemology and rational ethics, Vol. 15, pp. 137–158). London: Springer.

Parthemore, J. (2015). Specification of the unified conceptual space, for purposes of empirical investigation. In P. Gärdenfors & F. Zenker (Eds.), *Applications of conceptual spaces: The case for geometric knowledge representation* (pp. 223–244). Cham: Springer.
Parthemore, J. (2017). Consciousness, semiosis, and the unbinding problem. *Language & Cognition, 54*, 36–46.
Parthemore, J., & Whitby, B. (2012). Moral agency, moral responsibility, and artefacts. In D. J. Gunkel, J. J. Bryson, & S. Torrance (Eds.), *The machine question: AI, ethics and moral responsibility* (pp. 8–16). Society for the Study of Artificial Intelligence and Simulation of Behaviour (AISB). Available online from http://events.cs.bham.ac.uk/turing12/proceedings/14.pdf.
Parthemore, J., & Whitby, B. (2013). When is any agent a moral agent? Reflections on machine consciousness and moral agency. *International Journal of Machine Consciousness, 5*(2), 105–129.
Parthemore, J., & Whitby, B. (2014). Moral agency, moral responsibility, and artifacts: What existing artifacts fail to achieve (and why), and why they, nevertheless, can (and do!) make moral claims upon us. *International Journal of Machine Consciousness, 6*(2), 1–21.
Polanyi, M. (1958). *Personal knowledge: Towards a post-critical philosophy*. Chicago: University of Chicago Press.
Prinz, J. (2004). *Furnishing the mind: Concepts and their perceptual basis*. Cambridge, MA: MIT Press.
Quine, W. V. (1969). Natural kinds. In N. Rescher (Ed.), *Essays in honor of Carl G. Hempel: A tribute on the occasion of his sixty-fifth birthday* (pp. 5–23). Dordrecht: Springer.
Rosch, E. (1975). Family resemblances: Studies in the internal structure of categories. *Cognitive Psychology, 7*, 573–605.
Rosch, E. (1999). Principles of categorization. In E. Margolis & S. Laurence (Eds.), *Concepts: Core readings* (Chap. 8, pp. 189–206). Cambridge, MA: MIT Press.
Sharples, M. (1999). *How we write: An account of writing as creative design*. London: Routledge.
Stewart, J. (1995). Cognition = life: Implications for higher-level cognition. *Behavioural Processes, 35*(1–3), 311–326.
Torey, Z. (2009). *The crucible of consciousness: An integrated theory of mind and brain*. London: MIT Press.
Woodfield, A. (1994). Do your concepts develop? In C. Hookway & D. Peterson (Eds.), *Philosophy and cognitive science* (pp. 41–67). Cambridge: Cambridge University Press.
Zlatev, J. (2001). The epigenesis of meaning in human beings, and possibly in robots. *Minds and Machines, 11*, 155–195.
Zlatev, J. (2002). Meaning = life (+ culture). *Evolution of Communication, 4*(2), 253–296.

Chapter 7
Seeking for the Grasp: An Iterative Subdivision Model of Conceptualisation

Mauri Kaipainen and Antti Hautamäki

Abstract Concepts are fundamental collective constructs of individual items that are capable of abstracting meaningfully homogeneous groupings of phenomena. This capability is a prerequisite for communication and action and gives structure to learning and memory. Our study is aligned with the vast paradigm that assumes embodied cognition, rooted in Merleau-Ponty (Phenomenology of perception (trans: C. Smith). Routledge and Kegan Paul, London, 1962), seminally articulated by Varela et al. (Embodied mind: cognitive science and human experience. MIT Press, Cambridge, MA, 1991) and existing today in a number of variants that have been reviewed by Wilson (Six views of embodied cognition. Springer. Psychon Bull Rev 9(4):625–636, 2002). We argue that the faculty to conceptualise may spring from the ability of *homo habilis* to manage concrete actions in space and time, and we propose that at the root level, 'grasping concepts' in a cognitive perspective may have a lot to do with the process of 'grasping objects' from an operational position.

7.1 Introduction

In this chapter we theorise perspective-specific conceptualisation, or cognitive concept determination – as a sequence of mental actions that involves iteratively subdividing the universe of objects, and in the course of this process seeking for the perspective that best satisfies both the analytic intention and avoids conceptual redundancy. That is, conceptualisation involves analyses that reduce the number qualities (understood as dimensions of these objects) according to a particular order of priority that is based on the cognizer's intention, situation and context.

M. Kaipainen (✉)
Perspicamus LTD, Helsinki, Finland
e-mail: mauri.kaipainen@perspicamus.com

A. Hautamäki
Department of Social Sciences and Philosophy, University of Jyväskylä, Jyväskylä, Finland

M. Kaipainen et al. (eds.), *Conceptual Spaces: Elaborations and Applications*, Synthese Library 405, https://doi.org/10.1007/978-3-030-12800-5_7

Our formulation adopts Gärdenfors' Conceptual Spaces theory (CS) (2000) as its starting point. In this theory, concepts are elementary items grouped into convex clusters in an abstract space whose dimensions are constituted by the qualities of the items. According to Gärdenfors, "what convexity requires is that if two object locations x_1 and x_2 both satisfy a certain membership criterion [...] then all objects between x_1 and x_2 also satisfy the criterion" (2000, 71). That is, the objects belong to the same concept. Since its introduction, this theory has provided fertile soil for a vast range of theoretical and practical elaborations. For an overview, see e.g. Zenker and Gärdenfors (2015). We rely on CS as the basis for describing perspective-dependent analytic inquiry, a process we describe in terms of procedural dynamics. We adopt the idea of concepts as *clusters*: sets of objects that are more similar to each other than they are to objects outside. Clustering analysis is a family of statistical techniques, but following CS we consider it as a kind of model of the cognitive-perceptual function that contributes to conceptualization.

However, clustering in itself is not without its problems. Among other issues, here we focus on the fact that there are always many alternative potential ways to clustering the objects under analysis, and there are correspondingly multiple analytical readings. Thus, "clustering is in the eye of the beholder" (Estivill-Castro 2002). That is, due to the lack of objective criteria, the cognizer draws lines that delineate clusters based on whatever specific criteria they chose to employ; someone else, however, might make different subdivisions. Our approach adopts the perspectival augmentation of the conceptual spaces theory (Kaipainen and Hautamäki 2015) and elaborates it in a direction that examines clustering as a process in time. Simultaneously, we tackle the mentioned issues of clustering from an alternative angle, that is, by considering the formation of perspectives and the process of conceptualisation in terms of analyses that iteratively subdivide the domain of inquiry. Essentially this means breaking the overall nonlinear and multidimensional task of clustering into a sequence of linear mappings. This, in turn, implies hierarchical conceptualization.

The idea of *perspectives*, often also referred to as *points of view*, as subjective determinants of perception and cognition is both omnipresent in everyday cognition and language and an evergreen topic of psychology and philosophy. The notion of *perspectivism* is a form of pluralism both in metaphysics and epistemology that dates back to Leibniz. The idea later became associated strongly with Nietzsche, who in *Beyond Good and Evil* claimed that "there are no facts, only interpretations". Further, according to him, "one always knows or perceives or thinks about something from a particular perspective – not just a spatial viewpoint, but a particular context of surrounding impressions, influences, and ideas, conceived of through one's language and social upbringing and, ultimately, determined by virtually everything about oneself, one's psychophysical make-up, and one's history" (Solomon 1996, 195). As Baghramian puts it, there can be more than one correct account of how things are in any given domain (2004, Chapter 10). Neither is there any reason to expect that any given phenomenon will have just one 'correct' account. When we accept this, the issue becomes not which perspective is correct or true but rather how to explore and mutually relate multiple perspectives. Consequently, there is no need to assume that the exploration of perspectives is ultimately satisfied at some point, or to expect that perspectives will converge into any final or 'true' form.

We have long sought to formally elaborate the idea of perspectives. Already in 1986 Hautamäki had articulated the relation between points of view and conceptual spaces (see Hautamäki 1986, 2015, 2016). The contribution of our more recent work with respect to CS (Kaipainen and Hautamäki 2011, 2015, 2016) has been the distinction between *ontological space* (which we shorten to *ontospace*) and *perspectival space*.

The notion of *ontospace* corresponds technically to the conceptual space that Gärdenfors (2000) articulates, but in our model we assign it a different role. Ontospace is an abstraction of intersubjectively shaped conceptual space. It is not the space where concepts manifest themselves, but rather the one of *potential but uninstantiated conceptual structures*.

Perspectival space, in turn, relates to situational and context-specific conceptualisations of the ontospace. The intuition behind the distinction is that when one is talking about a certain domain, one must identify a set E of entities in this domain. A set of qualities $Q_1,....Q_n$ is used to describe these entities. Formally, the ontospace is defined as a Cartesian product of 'quality dimensions'. Quality dimension $<Q_i,D_i>$ consists of a quality (or variable) Q_i and a set D_i of values for that quality. If there are n qualities, then the ontospace is the product $\mathbf{A} = D_1 x...xD_n$, and its elements are n-tuples of the form $<d_1, \ldots, d_n>$, where each d_i belongs to the set D_i. Thus, all entities in the set E are represented by elements of the ontospace (for formal details see Hautamäki 2016). Although we are not discussing different kinds of qualities in this chapter, we are tempted to doubt whether the distinction between primary (length, form) and secondary (colours, taste) qualities is as clear as Locke was assuming. Note also that according to Kant all qualities are secondary (1783).

In our formulation, it is the perspectival space **B** where items of the ontospace become manifest as clusters in a *perspective-dependent* manner (Kaipainen and Hautamäki 2011, 2015). By definition, the dimensionality of the perspectival space is lower than that of the ontospace. Formally, the manifestation is due to the reduction function R_P from the ontospace A into the perspectival space B, where P refers to a perspective. R_P obeys perspective P, which we have defined as a prioritisation among the quality dimensions ('variables' in a mathematical or statistical context) that influences the apparent clustering. It can be regarded as an element corresponding to the interest, attitude or disposition of the cognizer towards the domain of inquiry. Metaphorically, a perspectival space can be imagined as a projection surface behind the ontospace in which the items inhabiting the ontospace appear clustered in different ways, depending on the angle of the light shone through it (Fig. 7.1).

Our model suggests how a number of one-dimensional perspectives (visualized as the torch beams in Fig. 7.1) can be integrated by means of prioritisation, which further implies a hierarchical system of perspectival spaces. This can be regarded as a way to represent a natural concept in a manner that indicates relations between

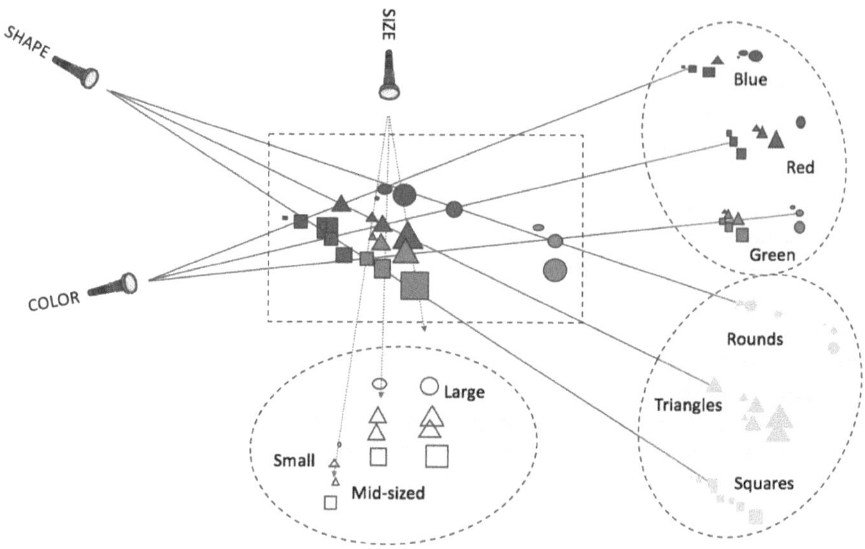

Fig. 7.1 An ontospace (dashed rectangle in the middle) inhabited by objects that are projected onto alternative perspectival spaces (depicted as dashed ovals), in which the clustering of items by similarity depends on the perspective adopted (depicted as torches)

different domains. It is, therefore, at least partially in accordance with criterion C (Gärdenfors 2011, 4) in terms of addressing the mereological relations among domains, that is, in terms of super- and subdomains.

However, even our perspectival elaboration of CS remains static in the sense that it does not describe the dynamical evolution of perspectives, concepts and spaces. In this chapter we further elaborate the notion of perspective-dependent concepts by describing an endlessly iterative model. For a thorough analysis for why concepts must evolve, as well as a comprehensive review of the discussion concerning their evolution, see Parthemore (2015, this issue).

Our approach regards concepts as transient constructs that emerge from continuous dynamic cognition. For a closer analysis of the dynamicity and the time scales in which concepts may evolve, see Parthemore (Ibid.). For now, we leave open the question of whether they are subjectively formed on the fly, as in perception, or whether they evolve more slowly as a function of situated and contextualized dispositions. We aim for a timescale-neutral model of how concepts—and perspectives that determine them—come about, and how they may guide 'next steps' in analytic cognition. In terms of the representation above (Fig. 7.1), our discussion here concerns the explorative sequence of analyses, altogether yielding to particular integrative concepts that serve to create cognitive 'graspability'. Before formalizing this iterative subdivision model of conceptualisation, however, we first relate the dynamics of conceptualisation to the embodied concretia of the action of hand and eye.

7.2 Grasping and Actionability of Concepts

Metaphorically, a perspective can be seen as an angle and a distance to an object of inquiry, allowing a view and therewith a 'grasp' of essential aspects of the object that would otherwise be obscured.

Reflected against on its Latin origin and relating it at least roughly to parallel expressions in many languages, the word *concept* suggests an association between the cognitive means and the manipulative action of the hand. The term derives from the Latin *con* (together) + *cept* (from *capere* = to capture, catch). *Con-* appears to hint at a plurality of primitive elements that can be grasped together at once. As another example, in the Finnish language the word for concept *käsite is* a composed from *käsi* ('hand') + *te* ('object to manipulate'), roughly *manipulatable*. In yet another language, the Swedish word for concept, *begrepp*, puts together the prefix *be-* connoting focus on a direct object and *grepp* (from *gripa* 'grasp'). These etymological associations suggest a longstanding link between the concrete action of a hand and the abstract notion of a *concept*. Both in light of the paradigm of embodied cognition that Varela et al. (1991) developed to address the bodily grounding of cognition, and in the spirit of cognitive metaphor theory from Lakoff and Johnson (1980), we suggest that the interpretation of *concept* as 'grasp' goes beyond mere etymological origins.

According to Cazeaux "a concept brings together what would otherwise be inchoate or diffuse into a unit, lump or sequence that makes recognition possible" (2017, 61). He recognizes the contribution of Merleau-Ponty (1962) as the inspiration for a whole range of discourse in philosophical and cognitive science on the embodied mind, encompassing what Sheets-Johnstone has called the "corporeal turn" (2015). At the epicentre of this would be the articulation of cognitive science as the study of the embodied mind, put forth by Varela et al. (1991) among others, along with Lakoff and Johnson's philosophy in the flesh (1999) and their earlier relation of metaphors to the bodily situation in the environment (1980). Gallese and Lakoff (2005) further argue for the neural grounds of such metaphors, referring to solid evidence that the sensory-motor system is able to model both sensory-motor and more abstract concepts, and that these modelings may not be separate but interwoven. Closely related to the embodied nature of cognition is the dynamic nature of cognition and the view of mind as motion, as discussed by van Gelder and Port (1995), among others. Equally relevant is the approach to cognition as a dynamic system (Smith and Thelen, 1994), and the recurrently dynamic perceptual cycle that Neisser suggested much earlier (1976). According to Neisser, a *schema* directs exploration, *exploration* samples the *object*, *object* modifies the *schema*, which again directs exploration, in an infinite loop (Fig. 7.2).

We do not have space here to exhaust the vast literature on the relationship between mind and body. We point the reader to Dawson (2013), who, among others, provides a broad picture of the interconnections between mind and body and world. Nevertheless, we believe that we are safe in proceeding with the idea of *grasping*

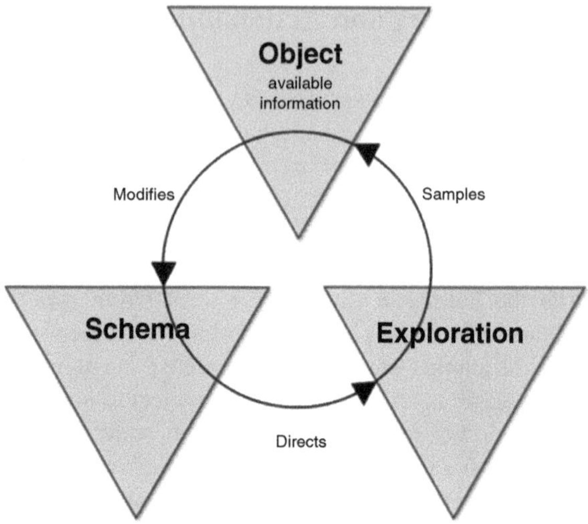

Fig. 7.2 Neisser's perceptual cycle (1976)

as a natural metaphor for 'what a concept is good for. We propose that this equals to making something accessible from an optimal working position, in a manner comparable to the work of a hand guided by vision.

The hand approaches objects from a particular angle and from a point of view that best orientates the concrete operation. Consider an agent engaged in concrete work, who seeks a perspective – or an operative position – in which the object's most critical dimension for successful operation is *as perpendicular as possible* to the eye's line of sight. This point of view then optimally reveals an angle as broad as possible and contributes to optimal precision in accessing the object (Fig. 7.3). In contrast, the operation of sawing a log would not be facilitated by, for example viewing the log from one if its ends and operating the saw traversely between left and right, which would be both visually difficult and ergonomically inefficient. Following the dynamic assumption in Neisser's perceptual cycle, we assume that the operative position and the optimal perspective are not given but rather result from learning through exploration (1976).

Prioritisation of qualities in perception and cognition is a classic topic across philosophy, psychology and cognitive science. For example, a round penny looks round only when held at a certain angle. From another perspective it appears to be a line, and from yet another an oval. In the case of colours, when one says that a rose is red, one presupposes standard lightning in which the rose appears to have this colour. Beside the perspective-bound nature of qualities, they are also connected to action, as C.I. Lewis pointed out (1956), proposing a temporal/causal relation between a quality and an action. According to Lewis, associating a quality with a thing implies acting in certain way vis-à-vis the thing, and as a consequence, specifiable experience will eventuate (ibid. 1956, 140). If I were to bite an apple, it would taste sweet. "The whole content of our knowledge of reality is the truth

Fig. 7.3 A lumberjack's operative position, aligning hand and vision perpendicular to the object to be cut. (Photograph by Anne LaBastille)

of such 'If-then' propositions in which the hypothesis is something we conceive could be made by our mode of acting" (ibid, 142). The consequence of if-then propositions is a possible experience caused by our action. Dewey, in turn, describes inquiry in a way that seems to refer to a progression of prioritisations such as the one we propose, such that it eventually turns into a holistic perspective: "[I]nquiry is the controlled or directed transformation of an indeterminate situation into one that is so determinate in its constituent distinctions and relations as to convert the elements of the original situation into a unified whole (1938, 108)." Further, he argues, "[thought] is a mode of directed overt action" (ibid., 166). Putting them all together, from Lewis and Dewey, we adopt the idea of *concept* as the result of a kind of 'recipe' (see also Parthemore 2015).

7.3 Iterative Subdivision Model of Conceptualisation

Our main suggestion is that perspective-specific hierarchical conceptualisation results from a sequence of iterative mental actions, which we compare to the operative positioning against concrete objects to be manipulated, as illustrated in Fig. 7.1. Further elaborating our earlier formulation (Kaipainen and Hautamäki 2015), where reduction function R_P was described as an abstract operation without specifying its temporal duration or sequence of steps, here we consider this transformation broken into a sequence of separate actions of subdividing clusters. This reasoning leads us to propose what we call the *iterative subdivision model of conceptualisation*. In this model there is a sequence in which one single dimension at a time is applied as the criterion for subdividing a cluster. In this sequence, clusters correspond to concepts that are determined in an increasingly focused manner, so that the items they stand for are optimally homogeneous with respect to each dimension chosen,[1] corresponding to what Dewey considered as "a unified whole" that is "determinate in its constituent distinctions and relations" (1938, 108). Thus, interpreting Dewey's "controlled or directed transformation" as a chain of operations (or a 'recipe'), in our formalisation we propose perspective P as the sequence of instructions on how to prioritise qualities, in order to arrive at a certain 'grasp' that makes a concept accessible from an intellectual 'operating position' with respect to the object of inquiry. In what follows we elaborate on this concept of perspective in further depth and illustrate how it works using a case example.

The reduction function R_P translates into a procedural sequence that unfolds over time that refine a perspective toward ontospace A. This sequence can be expressed as $P = [p_1,...,p_n]$, consisting of different real numbers from the set $\{1,2,...,n\}$, or from the set [0,1]. In this sequence, number p_i expresses the priority of the quality Q_i; the larger the number the lower the priority of the quality. Concerning the dependence of R_P on perspective P, it is stipulated that *the higher the priority of a dimension, the more globally it dominates the spatial organisation of the objects* due to its being executed earlier. Thus, if $p_i < p_j$, then quality Q_i has a more globally determinant role in organizing the spatial cluster's hierarchy than quality Q_j, and correspondingly the effects of dividing the cluster with Q_j as the criterion are more local. Regarding the highest level, we can suppose that priority p_g, constituting a 'genus-level' quality Q_g, is the one set at the highest level and is an implicit and self-evident presumption that determines the ontospace. In the example we will present later in this chapter, it precedes the conceptualisation process. For example, the genus-level quality is what constitutes the distinction between countries that are self-governed states in opposition to those that are non-states, such as mere geographical regions.

[1] Such clusters are also convex by nature, fulfilling the condition Gärdenfors stipulates (2000).

In the proposed approach, a perspective is built dynamically starting from Q_g, and thereafter searching for new qualities, taking into consideration those previously chosen, iteratively leading to the sequence of qualities [Q_g,Q*,Q**,...]. Any quality added to the sequence will result in further subdivision and contribute to an increasingly homogeneous local concept, which at the same time belongs to a broader conceptualisation based on the entire sequence.

In order to compare the described sequence of cognitive actions with the actions of a hand, we adopt the concept of *orthogonality* from linear algebra and statistics as an analog to the perpendicular positioning towards objects in concrete work. We stipulate that in conceptualisation, quality dimensions whose distribution is as orthogonal (statistically independent) as possible *of the superdimension*, (the dimension applied to determine the concept currently in focus), correspond to perpendicular working position against a concrete object to manipulate. Those are the distinguishing dimensions that are technically best suited for making iteratively finer distinctions among the elements of an ontospace, because it is not intellectually desirable that a distinguishing dimension should covary closely with the an already made distinction. This would make it redundant and it would fail to contribute new insights. The preference of nonredundant quality dimensions becomes quite intuitive if we look at an example from everyday classification. Maximal covariance (minimal orthogonality) between two qualities corresponds to synonymity: i.e. concepts that refer to more or less the same items. The opposite approach – choosing a maximally orthogonal subdividing dimension – is a method to avoid intellectually empty concepts. For example, it would not add any information to the concept of 'the bald' if we were to subdivide it further into 'hairless persons among the bald'. Thereagainst, if we were to pick another dimension that is more orthogonal with respect to 'baldness', say age, it would result in two potentially useful subconcepts, namely 'the young bald' and 'the old bald', a comparison between which might well be key to better understanding the concept of baldness.

However, maximal orthogonality alone does not guarantee that the distinction contributes to the practical analytical task that intention p aims to solve. Neither is it guaranteed that it reflects the continuously changing environment in a relevant way. We further suggest that in addition to orthogonality, the distinguishing dimension needs to contribute to the analytic intention – the reason the analysis is being undertaken in the first place and the motivation that drives the process throughout.

More generally, we stipulate that concepts are formed in a dialogue of inferential and intentional components, represented by statistical orthogonality and idiosyncratically situated intentionality, respectively. The optimisation of orthogonality and intentionality can be described in terms of a two-dimensional grid (Fig. 7.4).

In order to relate the two considerations – inferential and the intentional – to practical action, we adopt *practical syllogism* as a translational theoretical frame.

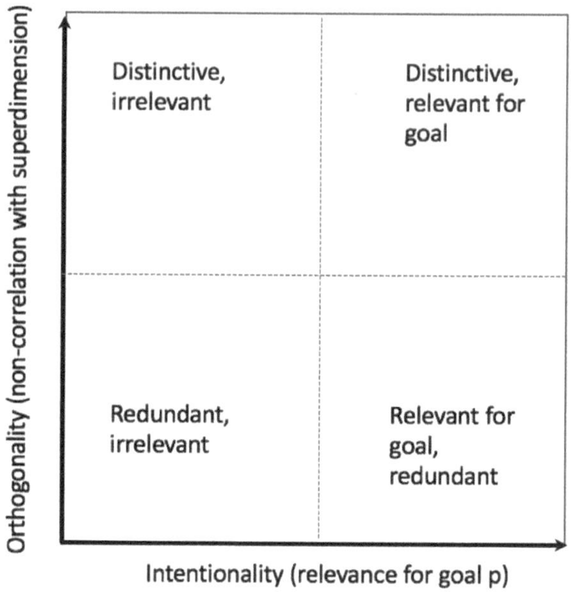

Fig. 7.4 Optimisation of intentional (relevance for goal on the x-axis) and inferential (distinctiveness vs. redundancy on the y-axis) considerations in the choice of subdividing criterion. The most desirable choice is as close as possible to the top-right corner

7.4 Practical Syllogism as the Recurrent Logic of Analytic Inquiry

Practical syllogism is a three-component logical argument with a major premise, a minor premise and a conclusion that takes the form of an action. It can be considered a logical means to bind together intention, knowledge (conceptual structure) and action. It differs from deductive inference, in which the result is a necessary consequence of the premises. Instead, the outcome of a practical syllogism is an option for action. One may also interpret the practical inference in terms of probability and propensity, as a tendency to act in certain conditions. We interpret this aspect in a dynamic manner, regarding the result of the syllogism as feedback for another iteration and involving choices among alternatives, altogether forming an open dynamic system directed towards a goal by means of intentions, as characterized by Fig. 7.4. Formally, there might be a sequence of actions $a_1,\ldots,a_k$, so that a_1 is the first step and a_k is the intended final step to reach the intended target. Von Wright's formulation of practical syllogism (1971, 96) serves this interpretation well. Here two premises imply a resulting action, as follows:

1. *A intends to bring about p*
2. *A considers that he cannot bring about p unless he does a.*
3. *Therefore agent A sets himself to do a.*

Below, we will draw a parallel with practical syllogism in von Wright's terms and what we see as the motivation for grasp-targeting cognitive action. We do this by means of a synopsis that compares the three components with the steps of the model algorithm (Table 7.1). Components 1–3 constitute the *intentional, inferential and practical* considerations involved at each iterative cycle of conceptual refinement. Here we suppose that the considerations in the table are related to a phase in iterative process $a_1,...,a_k$.

Formally, to describe the stipulated sequence as a whole, and a single action a_t within it, we adopt the following notation for the partial transformation. Let M be a sequence of quality dimensions from the list $Q_1,...,Q_n$. The order of quality dimensions in M also expresses the order in which the qualities are applied, so it can be said to establish a perspective. Let R_P be a reduction function. Then R/M refers to the restriction of R_P only to qualities in M. Let us say that $M = [Q_1,Q_3]$; then R/M takes into consideration only qualities Q_1 and Q_3, in that order. In the iterative process of conceptualisation, the projection from ontospace into perspectival space is elaborated stepwise in terms of a series of partial function $R/M_1, \ldots ,R/M_k$ where the set of relevant qualities in $M_1, \ldots ,M_k$ increases. At each step, the application of quality dimensions of M in their preferred order will produce a new sublevel in the hierarchical cluster structure. In terms of action theory, these categorisations are evaluated on the basis of the prospects they open for action (compare with premise 2 of practical syllogism).

In the next section we apply the abstract model described above to a concrete dataset in order to demonstrate the relevance of the model to a real-world situation using four iterations.

7.5 Case: Country-Level Indicators of Social Progress

We apply the above-described iterative procedure for subdivision to a dataset aggregated from a number of sources that describes indicators of social progress in various countries (Social Progress Imperative 2017), reflecting each step of von Wright's practical syllogism.

We assume bringing about new insights to poverty as the general analytic intention of the analysis *p*: more specifically, to determine whether there are qualities that distinctively characterise a concept (cluster) of countries in which satisfaction of basic human needs is low, such that it can be intervened on, that is, such that it allows some actionability. This might be motivated by a desire to improve living conditions in those countries.

Step 1

Intentional: The initial subdivision (action a_1) is determined by the above-mentioned intention alone, since there is no previously determined superdimension that would allow orthogonality heuristics to be calculated.

Inferential: The most natural choice for the subdivision criterion is quality dimension Q *Satisfaction of Basic_human_needs*.

Table 7.1 Synopsis comparing practical syllogism (von Wright 1971) (left) to stipulated cognitive action (middle) and steps of proposed algorithmic model (right)

	Practical syllogism (von Wright 1971, 96)	Aspects of cognitive distinction	Steps of model algorithm
1	*Intentional consideration. "A intends to bring about p."*	On the basis of a previous result, the cognizer:	The model assumes the cognizer's input;
		Evaluates whether intended p is to be maintained (in terms of Lewis, hypothesis confirmed), or whether the target has been reached.	-> End condition judgment
		Refocuses the intention to the sub-concept among those determined in the previous step (practical action) to be most relevant for the intended p.	-> Refocus decision
2	*Inferential consideration*: *"A considers that he cannot bring about p unless he does a."*	The cognizer makes hypothesis-forming inferences regarding the domain based on the current conceptualisation:	Model-algorithmic correspondent:
		The cognizer identifies the obstacle ("cannot grasp the intended p...") and consequent need for further analysis ("...without doing further analysis")	-> Assumed as input
		Considers (intuitively) the most distinctive options for action "a" for subdividing the current concept (best 'operative position')	The model forms a heuristic for the most distinctive variables by means of variables that are most orthogonal to the distinguishing dimension from the previous iteration (superdimension) within the cluster in focus (determined at step 1).
		Chooses, from among the best options for distinguishing dimensions, the one that should optimally serve the intention of grasping p.	-> Assumes the cognizer's choice
3	*Practical action*: *"Therefore agent A sets himself to do a."*	Chosen distinguishing dimension applied to concrete instances, resulting in new sub-concepts.	Algorithmic subdivision of the cluster in focus based on the distribution of elements on the chosen subdividing dimension (at step 2).

Practical: Applying perhaps the simplest possible clustering method, namely setting the widest gap in the distribution of countries described by the data set (Social Progress Imperative 2017) as the break point, the following cluster subdivision results:

Higher Satisfaction of Basic_Human_Needs:

Korea, Republic of Poland, Hungary, Greece, Italy,
Chile, Serbia, Bulgaria, Costa_Rica, Uruguay,
Albania, Turkey, Romania, Tunisia, Panama,
Thailand, Kazakhstan, Ukraine, Moldova, Russia,
Mexico, Ecuador, Kyrgyzstan, Morocco, Peru,
Brazil, Sri Lanka, Colombia, Bolivia, Indonesia,
Dominican Republic, Nicaragua, El Salvador, Philippines, Guatemala,
Botswana, Nepal, Honduras, Bangladesh, Mongolia,
India, Senegal, Pakistan, Cambodia, Ghana

Lower Satisfaction of Basic_Human_Needs:

Zimbabwe, Liberia Uganda Sierra Leone, Malawi,
Kenya, Cameroon, Côte d'Ivoire, Tanzania

This division can be considered as the highest-level conceptual structure, further constraining the analysis as it continues.

Step 2
Intentional: In the first iterative step (action a_1), the intention dictates that it is the cluster with *lower Satisfaction of Basic_Human_Needs* that is of interest, and consequently the exploratory focus of further analysis is narrowed to this cluster.

Inferential: The following inference then seeks to find an optimal 'operative position' for the analytic subdivision of cluster *Lower Satisfaction of Basic_Human_Needs.* According to our model, a spatial/cognitive position toward the ontospace, in mathematical terms, that is orthogonal to the superdimension generates hypotheses regarding which dimensions might make a distinction that is the least redundant and therefore can optimally contribute to an actionable concept. The top ten of 74 variables in the dataset (ibid.) ordered by descending orthogonality to the superdimension *Satisfaction of Basic_human_needs* lead to the following heuristic:

Primary_school_enrolment [choice]
Corruption
Tolerance_for_homosexuals
Freedom_of_religion
Press_Freedom_Index
Foundations_of_Wellbeing
Level_of_violent_crime
Suicide_rate
Household_air_pollution_attributable_deaths
Mobile_telephone_subscriptions_-_capped

In addition, assume that the cognizer forms a hypothesis, reminiscent of Lewis's suppositions, namely that the cluster of countries with a high average value on quality dimension (variable) Q* = *Primary_school_enrolment* might also be characterized by higher scores on other positive social progress indicators compared to those within the cluster with a lower average on this indicator. In terms of our formula, assume that this choice is M = [*Satisfaction of Basic_Human_Needs, Primary_school_enrolment*], equal to the new subdividing dimension Q*. The actionality criterion – a starting point of the analysis – might be one of the reasons to prioritise *Primary_school_enrolment* over, say, lowering *Corruption* or increasing *Tolerance_for_homosexuals,* because it may appear that policymakers may be able to take effective action in this dimension. Obviously, another cognizer might make another choice.

Practical: The correspondent epistemic action is to test the hypothesis concerning the practical benefit of the chosen Q*, presupposing cluster (concept) M, corresponding to *action* in practical syllogism. This is done by applying a dimension *Primary_school_enrolment* as the criterion to subdivide the cluster defined by the dimension *Lower- Satisfaction of Basic_Human_Needs*. Applying the method described in Iteration 1, the resulting subdivision occurs:

Lower- Satisfaction of Basic_Human_Needs:

Higher- Primary_school_enrolment:

Zimbabwe, Liberia, Uganda, Sierra Leone, Malawi, Kenya, Cameroon

Lower- Primary_school_enrolment:

Côte d'Ivoire, Tanzania

Step 3

Intentional: The issue to be evaluated is whether the subdivision by *Primary_school_enrolment* provides a sufficient 'grasp' on the domain of inquiry. This can be studied in terms of comparing the resulting clusters, scrutinizing how the variables covary within the supercluster in focus (chosen at Iteration 1) between the newly distinguished subclusters *Higher- Primary_school_enrolment,* and *Lower- Primary_school_enrolment.* In Fig. 7.5 the data variables are ordered by their ascending covariance with the selected dimension (which is equal to descending orthogonality). The column displays the comparable average values of respective variable (dimension) for each subcluster, scaled from 0 to 1 (Fig. 7.5).

These results may turn out to be rather surprising, or even counterintuitive. For example, we might not expect that the higher the *Primary_school_enrolment,* the higher the average *Maternal_mortality_rate,* but according to our data this holds within the cluster defined by *Lower- Satisfaction of Basic_Human_Needs.* This is indeed a new insight. In this case, the practical judgment may be that *Primary_school_enrolment* does not suffice to provide the intended grasp of the domain, and a more refined analysis is needed. As before, the model is not intended to, and cannot, explain the human choice of the analytic variable. Someone else's

	Lower	Highe	
Primary_school_enrollment	0.202	**0.709**	**GIVEN**
Deaths_from_infectious_diseases	0.754	**0.83(**	**Unexpected**
Early_marriage	0.056	**0.059**	
Access_to_Basic_Knowledge	0.152	**0.419**	
Maternal_mortality_rate	0.382	**0.483**	**Unexpected**
Personal_Safety	0.436	**0.467**	
Child_mortality_rate	0.578	**0.605**	**Unexpected**
Homicide_rate	0.130	**0.065**	
Household_air_pollution_attributable_(	0.748	**0.721**	
Tolerance_and_Inclusion	0.372	**0.345**	
Secondary_school_enrollment	0.115	**0.194**	
Secondary_school_enrollment_-_capp	0.162	**0.272**	
Globally_ranked_universities	0.000	**0.005**	
Globally_ranked_universities_-_buckel	0.000	**0.001**	
Percentage_of_tertiary_students_enro	0.000	**0.143**	
Access_to_improved_sanitation_faciliti	0.067	**0.182**	
Freedom_of_expression	0.241	**0.241**	
Greenhouse_gas_emissions	0.389	**0.549**	**Unexpected**
Greenhouse_gas_emissions_-_cappe(	0.374	**0.538**	
Level_of_violent_crime	0.651	**0.601**	
Quality_of_electricity_supply	0.014	**0.009**	

Fig. 7.5 Comparison of subclusters *Higher- Primary_school_enrolment* and *Lower- Primary_school_enrolment,* in terms of average values of variables (scaled from 1 to 0). The authors' comments are added on the right

conclusion might be to reject this variable and go back and try another one. The result could also be interpreted as an impetus for setting some hypothesis for empirical studies: for example, to test whether there is a causal relation between higher *Primary_school_enrolment* and higher *Deaths_from_infections_diseases*. Perhaps contagious diseases are spread through schools?

Assume, however, that the cognizer makes the decision to focus on countries within the cluster with higher (rather than those with lower) average values for *Primary_school_enrolment* in the next step. This could be motivated by the unexpected finding of higher *Maternal_mortality_rate* in this cluster, implying the need for further study. Better clarity on this issue should serve the overall analytic intention p, namely to gain new insights into poverty.

Inferential: As in step 1, the variables are ordered by descending orthogonality to superdimension *Primary_school_enrolment* as a heuristic for next cluster subdividing dimension.

Opportunity
Access_to_electricity
Biodiversity_and_habitat
Access_to_Information_and_Communications
Press_Freedom_Index
Depth_of_food_deficit
Rural_access_to_improved_water_source
Depth_of_food_deficit__capped
Mobile_telephone_subscriptions
Mobile_telephone_subscriptions__capped
Access_to_piped_water [choice]

In this case, let us assume the dimension (variable) *Access_to_piped_water* is chosen from among the alternatives. The choice might be based on the considerations following from the unexpected finding at step 2: namely, to investigation which variables might contribute to the seemingly unlikely positive covariance of *Deaths_from_infectious_diseases* with *Primary_school_enrolment*.

In this context, *Access_to_piped_water* seems worth exploring. In addition, access to piping is not just an observable phenomenon but it refers to an actionable measure that can improve the overall quality of life in the countries covered by the concept.

Practical: Using the dimension *Access_to_piped_water* as the new criterion Q** to subdivide *Higher- Primary_school_enrolment* yields the following subdivision hierarchy:

Higher- Primary_school_enrolment:

Higher- Access_to_piped_water:

Zimbabwe, Kenya, Cameroon

Lower- Access_to_piped_water:

Malawi, Sierra Leone, Uganda, Liberia

Step 4
Intentional: Variables that covary with *Access_to_piped_water* between the new subclusters within the supercluster [*Lower- Satisfaction of Basic_Human_Needs, Higher- Primary_school_enrolment*] can be scrutinized with a comparative table of the kind we used earlier (Fig. 7.6).

As the result, the sequence [*Satisfaction of Basic_human_needs, Primary_school_enrolment, Access_to_piped_water*] constitutes perspective P that defines a particular concept, referring to a group of countries that are homogeneous from a particular perspective, in turn determined by the analytic intention. In this perspective, a cluster appears that corresponds to a concept that might be called *"Tap countries"* based on its most local distinction, taking the higher-level distinctions

	Lower	Higher	
Access_to_piped_water	0.028	0.200	**CRITERION**
Women's_average_years_in_s	0.201	**0.537**	**Superior**
Adult_literacy_rate	0.534	**0.774**	**Superior**
Adult_literacy_rate_-_capped	0.560	**0.791**	**Superior**
Availability_of_affordable_hou	0.001	**0.002**	
Shelter	0.113	**0.294**	
Gender_parity_in_secondary_	0.002	**0.002**	
Access_to_Basic_Knowledge	0.317	**0.555**	**Superior**
Political_terror	0.393	**0.714**	**Arising issue**
Secondary_school_enrollment	0.110	**0.305**	**Superior**
Secondary_school_enrollment	0.154	**0.429**	
Perceived_criminality	0.500	**0.833**	**Arising issue**
Access_to_electricity	0.125	**0.387**	
Satisfied_demand_for_contrac	0.536	**0.738**	**Superior**
Press_Freedom_Index	0.245	**0.334**	
Suicide_rate	0.043	**0.076**	
Internet_users	0.058	**0.216**	
Access_to_Advanced_Educati	0.127	**0.290**	**Superior**
Access_to_improved_sanitatio	0.107	**0.281**	
Foundations_of_Wellbeing	0.112	**0.199**	
Access_to_Information_and_C	0.159	**0.294**	
Religious_tolerance	0.833	**0.667**	**Arising issue**
Water_and_Sanitation	0.164	**0.225**	
Deaths_from_infectious_disea	0.819	**0.859**	**Remaining issue**

Fig. 7.6 Comparison of selected average values of variables between the new subclusters *Higher-* and *Lower- Higher- Access_to_piped_water* within cluster *Primary_school_enrolment,* with authors' comments on the right. Font size denotes the value, while bold indicates the choice of focus

for granted. The concept (underlined) and its hierarchical context ar indicated by the hierarchical outline below.

Higher- Satisfaction of Basic_Human_Needs:

. . .

Lower- Satisfaction of Basic_Human_Needs:

Higher- Primary_school_enrolment:

Higher- Access_to_piped_water:

Zimbabwe, Kenya, Cameroon <- "Tap countries"

Lower- Access_to_piped_water:

Malawi, Sierra Leone, Uganda, Liberia

Lower- Primary_school_enrolment:

Côte d'Ivoire, Tanzania

In terms of language, this concept may be conveniently described in the reverse order as the one referring to countries with

High- *Access to piped_water ("Tap countries"),* among those with
High- *Elementary_School_enrolment,* among all those with
Low- *Satisfaction of Basic_Human Needs.*

We conclude by pointing out that this concept allows an actionable 'grasp' in terms of the comparison between Low- vs. High *Access_to_piped_water*, as it turns out that this distinction is quite crucial. The ones in the previous subcluster are superior to the latter with respect to many significant average dimensions: for example *Maternal_mortality_rate, Depth_of_food_deficit* and *Child_mortality_rate*. Identifying this concept constitutes a rather practical aspect of 'grasping', implying, for example, water infrastructure improvements versus *"Without-tap-countries"*. The hierarchical structure it belongs to represents the understanding of the conditions in which this concept is applicable. It may suffice to point out intuitively that the cognizer is often only implicitly or unconsciously aware of a concept's conditions of applicability. The top-down hierarchical analysis we propose, however, makes it explicit.

Altogether, it may seem justified for the cognizer to deem that the intended goal p is met and to terminate the iteration, if not finally at least for the time being. In fact, this approach does not assume that a final conceptualisation exists at all. Rather, further subdivisions might be needed to bring clarity to the newly arising local correlations or to test hypotheses that may occur to the cognizer.

7.6 Discussion and Conclusion

Our chapter is an attempt to describe the cognitive dynamics of how concept-defining perspectives come about, and how they determine perspective-relative conceptual structures. Building on philosophical, logical and cognitive science arguments, we suggest that it takes place by means of analysis of conceptualisation through *iterative subdivision*. Our approach builds on three theoretical pillars.

First, the theory of Conceptual Spaces (Gärdenfors 2000) constitutes the fundamental spatial metaphor for describing concepts as spatially concentrated clusters of homogeneous items. Our perspectival interpretation of this theory (Kaipainen and Hautamäki 2015) offers the means to describe analytical cognition as conditioned by perspectives – that is, foci and choice of dimensions that correspond to situations, contexts and conditions of human activity.

Second, the bodily grounding of cognition and experience (for example Merleau-Ponty 1962; Varela et al. 1991; Lakoff and Johnsson 1980, 1999; Gallese and Lakoff 2005; Rizzolatti and Arbib 1998; Sheets-Johnstone 2015), in harmony with the association of cognition with action, motion and dynamics (Neisser 1976; Smith and Thelen 1994; Van Gelder and Port 1995) forms the justification for comparing concept-forming exploration to the concrete interplay of hand and eye. From this paradigm we derive the metaphor of *concept as grasp*, as well as that of *operative positioning* with respect to an object to be manipulated. The core suggestion we put forward is that exploratory sequences of concept-building actions may obey constraints just like the actions of the hand in the physical world: Just as a *perpendicular* orientation toward a physical object is often the most ergonomic and accurate position from which to manipulate it, a position that is orthogonal to already-assumed distinguishing quality dimensions can usefully guide the iterative construction of concepts.

As the third foundational element of theory, we have employed von Wright's articulation of *practical syllogism* (1971) as the structure for our iterative sub-division model and as a means of translating our perspectival conceptual space model into a temporal sequence. In addition, this approach addresses, though it may not exhaust, the question lurking behind the perspectival explanation: How do conceptualisation-defining perspectives come about? On the other hand, our iterative model allows us to propose how practical syllogism might feed back into itself, implying a dynamic evolution of intentions, and how intentions may drive perspective-searching exploration. This circularity is inherited from pragmatism and also shows a clear resemblance to Neisser's perceptual cycle, discussed earlier.

However, we accept von Wright's warning (1971, 95; 97) against interpreting practical syllogism as causal inference, because the complexity of the cognitive-volitional component does not qualify as a cause, at least in the Humean sense. Our model may be explicit when it comes to inferential considerations and practical action, but it would be a false simplification to interpret it as a causal or deterministic explanation, due to the open-endedness of intentionality. It is difficult to imagine a formal model, let alone an algorithm, that exhausts bodily implementation, ergonomics, situatedness, life stories, and social setting – among other factors – that would *causally* motivate intention and will. It is tempting to predict that even the ultimate limit of artificial intelligence may eventually be determined by the same issue: that is, how can a machine, even in principle, mimic such complexity?

We have proposed how iterative sub-clustering refines concepts so that they correspond to the analytic intention. However, none of the resulting concepts needs to be final; rather, it is natural to assume that inquiry will proceed in terms of erasing prior divisions and forming new subdivisions using alternative criteria. The exploration process may continue ad infinitum, forming an endless track of trials, some of which may turn out to be mistaken, while others may contribute to a broader conceptualisation.

How separate explorations form the big picture is another issue, and it remains clearly beyond the explanatory capabilities of present model. Two alternative suggestions seem worth considering: (1) The cognizer is simply unable to see the full picture at all, or (2) some kind of temporal Gestalt emerges connecting the 'dots', perhaps in terms of a story about the course of exploration told to oneself or to others. Elsewhere, there is an ongoing line of empirical studies that assume there may be neural systems that generalise over memories of related events, perhaps such as those that Schlichtning et al. (2015) have observed in the hippocampus and prefrontal areas. However, we suspect that the prerequisites for a comprehensive cognitive model that integrates conceptual explorations across time are not sufficient at this time, either in the cognitive sciences or in any other empirical field.

To conclude, while we may have failed to exhaust intentionality in algorithmic terms, we nevertheless suggest a cognitive-computational dynamics of interaction as a way to connect the categorisation-oriented theory of conceptual spaces to explorational perspectival conceptualisation. Essentially, in seeking a way to relate concepts to unfolding action, the proposed model moves beyond the purely spatial metaphor of conceptualisation toward a spatiotemporal description.

Acknowledgements Writing this chapter was financially supported by The Foundation for Baltic and East European Studies (Östersjöstiftelsen), Sweden.

References

Baghramian, M. (2004). *Relativism*. London/New York: Routledge.

Bastille, A. *Photograph of a lumberjack*. Public domain, NARA record: 1422473, Wikimedia Commons.

Cazeaux, C. (2017). *Art, research, philosophy*. London: Taylor & Francis.

Dawson, M. R. W. (2013). *Mind, body, world: Foundations of cognitive science*. Edmonton: Athabasca University Press.

Dewey, J. (1929). *The quest for certainty*. New York: Minton, Balch & Co.

Dewey, J. (1938). *Logic, the theory of inquiry*. New York: Henry Holt.

Estivill-Castro, V. (2002). *Why so many clustering algorithms — A position paper*. ACM SIGKDD Explorations Newsletter.

Gallese, V., & Lakoff, G. (2005). The brain's concepts: The role of the sensory-motor system in conceptual knowledge. Taylor & Francis. *Cognitive Neuropsychology, 22*(3–4), 455–479.

Gärdenfors, P. (2000). *Conceptual spaces: On the geometry of thought*. Cambridge, MA: The MIT Press.

Gärdenfors, P. (2011). Semantics based on conceptual spaces. In *Logic and its applications* (pp. 1–11).

Hautamäki, A. (1986). *Points of view and their logical analysis*. Helsinki: Acta Philosophica Fennica.

Hautamäki, A. (2015). Change, event, and temporal points of view. In M. V. Campos & A. M. L. Gutiérrez (Eds.), *Temporal points of view, subjective and objective aspects* (pp. 197–221). Heidelberg/New York/Dordrecht/London: Springer.

Hautamäki, A. (2016). Points of view: A conceptual space approach. *Foundations of Science, 21*(3), 493–510.

Kaipainen, M., & Hautamäki, A. (2011). Epistemic pluralism and multi-perspective knowledge organization. Explorative conceptualization of topical content domains. *Knowledge Organization, 38*(6), 503–514.
Kaipainen, M., & Hautamäki, A. (2015). A perspectivist approach to conceptual spaces. In F. Zenker & P. Gärdenfors (Eds.), *Applications of conceptual spaces: The case for geometric knowledge representation* (Synthese Library Volume 359). Dordrecht: Springer.
Kaipainen, M., & Hautamäki, A. (2016). Analysis and synthesis with a three-component inferential system: Augmenting the explanatory scope of Conceptual Spaces. *AIC 4th international workshop on artificial intelligence and cognition*. New York City, July 16–17 2016. Accessible at http://ceur-ws.org/Vol-1895/paper11.pdf
Kant, I. (1783). *Prolegomena to any future metaphysics that will be able to present itself as a Science* (orig. in German).
Lakoff, G., & Johnson, M. (1980). *Metaphors we live by*. Chicago: The University of Chicago Press.
Lakoff, G., & Johnson, M. (1999). *Philosophy in the flesh: The embodied mind and its challenge to Western thought*. New York: Basic Books.
Lewis, C. I. (1956). *Mind and the world order, outline of a theory of knowledge*. New York: Dover.
Merleau-Ponty, M. (1962). *Phenomenology of perception* (Colin Smith, Trans.). London: Routledge and Kegan Paul.
Neisser, U. (1976). *Cognition and reality. Principles and implications of cognitive psychology*. San Francisco: W. H. Freeman and company.
Parthemore, J. (2015). Specification of the unified conceptual space, for purposes of empirical investigation. Springer. In F. Zenker & P. Gärdenfors (Eds.), *Applications of conceptual spaces: The case for geometric knowledge representation* (Synthese library volume 359) (pp. 223–244). Dordrecht: Springer.
Rizzolatti, G., & Arbib, M. A. (1998). Language within our grasp. Elsevier. *Trends in Neurosciences, 21*(5), 188–194.
Schlichting, M. L., Mumford, J. A., & Preston, A. R. (2015). Learning-related representational changes reveal dissociable integration and separation signatures in the hippocampus and prefrontal cortex. *Nature Communications, 6*, 8151.
Sheets-Johnstone, M. (2015). *The corporeal turn: An interdisciplinary reader*. Andrews UK Limited.
Smith, L. B., & Thelen, E. (Eds.). (1994). *A dynamic systems approach to the development of cognition and action*. Cambridge, MA: MIT Press (A Bradford Book).
Social Progress Imperative. (2017). *Social progress index*. Data set. accessed in September 2017 from http://www.socialprogressindex.com/?tab=4
Solomon, R. C. (1996). Nietzsche ad hominem: Perspectivism, personality and ressentiment. In B. Magnus & K. M. Higgins (Eds.), *The Cambridge companion to Nietzsche* (pp. 180–222). Cambridge: Cambridge University Press.
Van Gelder, T., & Port, R. F (1995). It's about time: An overview of the dynamical approach to cognition. In *Mind as motion: Explorations in the dynamics of cognition* (Vol. 1, p. 43). Cambridge, MA: MIT Press.
Varela, F., Thompson, E., & Rosch, E. (1991). *Embodied mind: Cognitive science and human experience*. Cambridge, MA: MIT Press.
Von Wright, G. (1971). *Explanation and understanding*. Ithaca: Cornell University Press.
Zenker F., & Gärdenfors P. (2015). *Applications of conceptual spaces* (Studies in Epistemology, Logic, Methodology, and Philosophy of Science, Vol. 359). Springer. Synthese library.

Part III
Concepts and Disciplines

Chapter 8
Lost in Space and Time: A Quest for Conceptual Spaces in Physics

Sylvia Wenmackers

Abstract In this chapter, I investigate whether dimensions in physics are analogous to quality dimensions (in the sense of Gärdenfors P, Conceptual spaces: the geometry of thought, 2nd edn. MIT Press, Cambridge, MA, 2000, 2004) and whether phase spaces are to be considered as conceptual spaces (as proposed by Masterton G, Zenker F, Gärdenfors P, Eur J Philos Sci 7:127–150, 2017). To this end, I focus on the domain of force in classical physics and on the dimension of time from classical to relativistic physics. Meanwhile, I comment on the development of abstract spaces with non-spatial dimensions, such as conceptual spaces, which is itself part of a long history of conceptual development.

Keywords Conceptual space · Dimensional analysis · Phase space · Philosophy of physics · Relativity theory · Time

8.1 From the Concept of Physical Space to Other Conceptual Spaces

Recently, geometry has been applied to represent human concepts, both psychological (or phenomenal) concepts and scientific (or theoretic) ones (Gärdenfors 2000, 2004). This seminal work contains some suggestive comments on conceptual spaces (CSs) in physics, followed-up by Gärdenfors and Zenker (2013) and an interesting case study recently presented by Masterton et al. (2017). These authors proposed to apply CSs as a means to study conceptual changes in science: they show intermediate, gradual steps in what is usually regarded as a revolutionary shift.

Using geometrical dimensions for anything other than representing the three spatial dimensions was alien to Greek mathematicians, including Ptolemy and

S. Wenmackers (✉)
Centre for Logic and Philosophy of Science (CLPS), Institute of Philosophy, KU Leuven, Leuven, Belgium
e-mail: sylvia.wenmackers@kuleuven.be

M. Kaipainen et al. (eds.), *Conceptual Spaces: Elaborations and Applications*, Synthese Library 405, https://doi.org/10.1007/978-3-030-12800-5_8

Pappus (Macagno 1971: 391). In the post-Cartesian algebraic description of spatial relations and forms, it became as easy to write x^3 as x^4 or higher powers. And in the nineteenth century, physicists like Fourier (1822), Weber, and Gauss started using dimensions for non-spatial quantities, as synthesized by Maxwell (Roche 1998). So, the introduction of CSs can itself be regarded as constituting a step towards a nett radical shift in Western thought, broadening the notion of 'space' from the concrete physical to the ever more abstract. It is part of the process of modern science to construct abstract representations such as CSs, because they aid in concisely describing, explaining, and predicting phenomena.

My aim in the current chapter is to evaluate whether there are any spaces in physics that play a role for scientific concepts that is sufficiently close to CSs for psychological concepts to designate them as such. Of particular interest are phenomenal CSs that attempt to be perceptually uniform, such that the same average perceived difference (or dissimilarity) between pairs is represented by equal distance in the space. They are based on similarity judgments by subjects, who typically do not explicitly consider the relations between hues in geometrical terms. Throughout this paper, I consider the CS for the phenomenal concept of colour as a guiding example.[1] There are multiple candidates CSs for colour, each with three integral quality dimensions that allow us to represent all humanly perceptible hues. In colour CSs that attempt to capture perceptual uniformity, hues that receive the same name in a given natural language are found to correspond to a convex volume. Gärdenfors's criterion P (2000, 2004: 71) makes explicit the assumption (shared by prototype theory) that natural properties ought to be convex in this sense (where properties are regarded as a special kind of concepts; see 2000, 2004: 60).

It is not clear upfront which parts of physics CSs should represent: do they cover branches of physics, particular theories therein or individual hypotheses? I want to leave this open at the start of this exploration. Therefore, I first consider a number of concepts in physics and compare them to the kind of concepts for which CSs have already been developed; in particular, I consider phase diagrams as candidate CSs (Sect. 8.2.1). Then, I turn to the use of dimensions in physics, with special attention to the practice of dimensional analysis; considering possible spaces spanned by these dimensions, leads me to pose three questions (Sect. 8.2.2). Next, I take stock of some abstract spaces used in physics, evaluating which one of those – if any – can be interpreted as CSs (Sect. 8.2). To address some of the earlier questions in more detail, I focus on force in classical physics (Sect. 8.3) and on time in classical and relativistic physics (Sect. 8.4). The exploration ends with a discussion (Sect. 8.5).

[1] I use this example because I know it best (Douven et al. 2017), although I acknowledge that a drawback of relying on a single example is that it might be atypical in some ways.

8.2 Identifying Conceptual Spaces in Physics

The notions of physical dimensions and abstract spaces are all-pervasive in physics and they suggest a straightforward analogy to quality dimensions and CSs. Our task here is to check whether there indeed are viable candidates and, if so, what is the benefit of the reinterpretation for the philosophical study of scientific concepts. This approach implies that we are not primarily looking for an explicit representation of how physicists implicitly use their concepts, which is different from the work on psychological concepts (but see Sect. 8.5).

In addition, the colour CS does not represent the concept colour in a space among other concepts; instead, it represents hues, which can be grouped into particular colour concepts (in plural; such as blue, green, and orange) as volumes within that space. So, it is not to be expected that the CS formalism will allow us to track the shape of fundamental concepts used in physics across time. This would require a CS for concepts and possibly a hierarchy of concepts, which goes beyond the scope of the current exploration. Still, reconstructing CSs can be relevant for studying changes to a given concept over time, as Masterton et al. (2017) demonstrated. We should also be aware that this may not suffice to study all physical concepts, as we will see now.

8.2.1 Phase Diagrams as Candidate CSs

Concepts in physics are very diverse. Just to get a flavour of the variety, consider these twelve concepts: temperature and heat, interference and diffraction, fermion and boson, gauge and symmetry-breaking, electric field and quantum field, phase transition and critical point.[2] It is not guaranteed that all central concepts in a scientific field will or even should be amenable to an analysis or a representation in terms of CSs. While we will see below that temperature is a dimension (and a domain) of a CS, it is not clear that all concepts on this list can be connected to a CS (in some restricted number of ways).

The main question in this section is whether any of the mentioned concepts are sufficiently analogous to concepts outside physics for which a CS has been specified already.

Gärdenfors (2000, 2004) stipulates that a point in a CS represents a possible object: this suggests that few of the initial concepts are of the right kind to expect finding a readymade CS in physics. By the looks of it, phase diagrams are good candidates.

Phase diagrams are graphs that represent phases of matter in equilibrium at particular combinations of thermodynamic variables (see Fig. 8.1 for a generic

[2] Some of the concepts are also central to other natural and applied sciences, such as chemistry and engineering. We will consider physics as a common ground for all examples in this chapter.

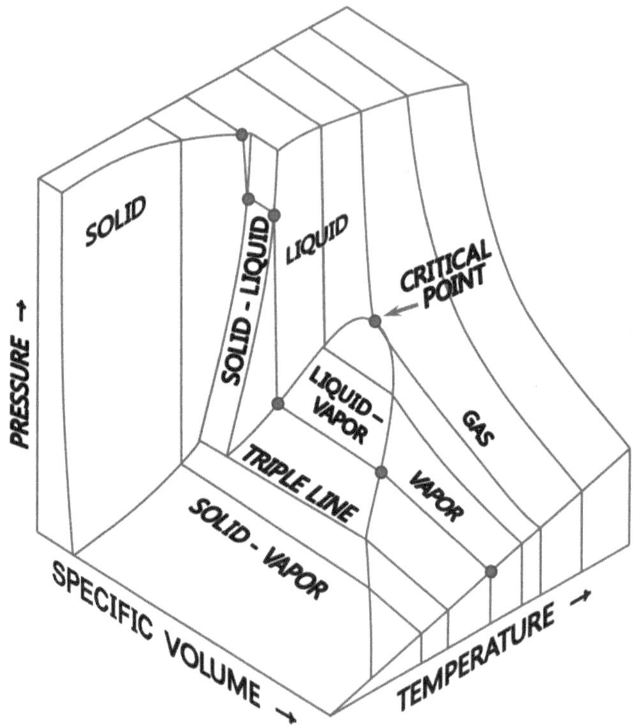

Fig. 8.1 Three-dimensional phase diagram showing equilibrium phases of some pure substance for particular combinations of temperature, pressure, and specific volume. (Image from a 1953 textbook by D.L. Smith, traced by Kyu Ho Lee and revised by H. Padleckas)

example). Given a substance (*e.g.*, water), the phase diagram shows all possible combinations of thermodynamic variables for a sample of it and the continuous regions (volumes) represent phases in which the substance may exist (*e.g.*, gas, liquid, and various solid phases). So, like a CS, a phase diagram can be seen as a representation of possible objects (or systems).

The axes of a phase diagram represent possible values of thermodynamic variables, which can include temperature, pressure, specific volume, specific enthalpy, and specific entropy. In the case of substances that are mixtures, additional axes can represent one or more variables for the composition (such as concentration). These can be interpreted as quality dimensions. So, in general, these are high-dimensional spaces, although in practice one often encounters two- or three-dimensional representations, sometimes parametrized with further variables (iso-lines).

Also phase transitions and critical points can be represented by lines and points in a phase diagram, but such a diagram is *not* used as a space to represent all possible phase transitions or critical points (occurring across different materials) – even though it would have suitable dimensions for the latter. So, a phase diagram is not the space on which these concepts 'live'.[3] Concepts that have been developed initially for classical states of matter have been generalized to more abstract states. Reasoning by analogy is important in physics; a well-known example is Maxwell's work on electromagnetism from analogy to fluid mechanics. Searches for mappings between two domains can lead to conceptual change in both domains (Carey and Spelke 1994). So, some concepts don't live in the domains, but are preserved across mappings between them.

We can now reverse-engineer the example of phase diagrams for our purposes: if it is to be considered as a CS, what is the concept it represents? Let us consider the CS for colour for guidance: *that* CS represents hues according to multiple quality dimensions, and named colours are found as convex regions of the space. Although particular phases in a phase-diagram are not necessarily convex (see the liquid phase near the critical point in Fig. 8.1) and they 'touch' rather than leaving out undesignated parts of the space, they do seem to be the obvious analogue to particular colours in the colour CS. So, if we follow the analogy through, the phase diagram may be interpreted as a CS for the concept 'phase of matter of a given material'.[4] Particular points in the CS represent particular thermodynamic conditions in which a sample of the material can exist, whereas the regions separated by phase boundaries (on which certain phases can exist in equilibrium) correspond to particular stable phases of the material. In fact, with this example we remain close to a perceptual CS, since many phases of matter are easily discriminable by the senses (just like specific colours are), although in other cases scientific instruments are needed (*e.g.*, to identify various crystalline phases). The main difference is the origin of the similarity measure, which is perceptual for colours, but measured (albeit by macro-variables) for a phase space.[5]

So, we see that if we consider phase diagrams to be CSs, their multiplicity makes them quite different from, say, the colour CS. In any case, phase diagrams are not the CS for states or phases of matter as such, which can be modelled (as in footnote 3), but which is not used in science. As this example also shows, some central concepts in physics, such as temperature, are physical dimensions: we will elaborate on this in the next section, to see whether all such physical dimensions should be thought of

[3]As suggested by a referee, the concept 'classical states of matter' could be modelled by a CS as follows: a discrete Euclidean space with values 0 and 1 along dimensions solid, liquid, gas, and plasma. Equilibrium states and critical points would be indicated by vectors that are not aligned with a particular axis. It is possible to distinguish more states (such as amorphous or crystalline solid) and phases thereof (although that becomes a very long list if the material isn't specified upfront), or to include modern states (such as Bose-Einstein condensate) and very high energy states (such as quark-gluon plasma), which leads to a higher dimensional CS.

[4]Hence, a possible disanalogy between phase diagrams and colour CS is that the same dimensions can be used to represent phases of different materials. This seems different from the way 'the' CS of colour is discussed in the literature, although we could in principle consider multiple colour CSs to distinguish, *e.g.*, between different human individuals or groups of humans belonging to different language communities, *etc.*

[5]We can feel temperature and pressure, but our crude bodily perception of them does not factor into the organization of the phase diagram.

as axes of CSs. In Sect. 8.2.3, we will see how to combine dimensions into spaces: there, we will consider phase diagrams as a first example of a broader class of spaces used in physics.

8.2.2 Dimensioned Quantities in Physics

It is well known – yet easily forgotten – that Newton never wrote his second law as '$\mathbf{F} = m\mathbf{a}$' (where boldface indicates vector quantities and $\mathbf{a} = d^2\mathbf{r}/dt^2$). He expressed this axiom in words,[6] not using his calculus,[7] and real-valued functions and vectors were only developed later on. Newton elaborated on his second law using ratios and proportions of the relevant quantities (using Euclid's theory of proportions), which is similar to the notation Galileo used and which implicitly encodes the comparative aspect of any physical measurement; similarly for other passages in the *Principia*.

Since we now use the International System of Units (SI) as a universally accepted conventional choice for units, we can leave these ratios implicit. Admittedly, implicit use of ratios was used in analytic geometry before the introduction of SI (see footnote 10) and Euler did use equations in terms of absolute rather than relative dimensioned quantities prior to the establishment of SI (Schulman 2010), but the introduction and rapidly widely adopted system of units has certainly helped the propagation of equations of this form (*e.g.*, $F_x = ma_x$).

In our search for an analogon to quality dimensions and CSs, the practice of dimensional analysis (DA) can serve as a source of inspiration. As we will see below, physical dimensions express the quality of various physical quantities.

Fourier (1822) introduced the concept of physical dimension[8] in his treatise on the analytical theory of heat. DA has since been used in various branches of physics[9] and related fields (in particular, engineering), especially in their didactics. The best known application of DA is to check the homogeneity of equations in physics: both sides of an equation (or inequality) should have the same dimensions, and hence the same units; likewise for terms in a sum.[10] But DA is much richer: it provides a way to determine the general form of relations that apply between physical quantities in a particular system, even when the special laws are not known yet.

---

[6]Newton (1697, 1999: 416): "A change in motion is proportional to the motive force impressed and takes place along the straight line in which that force is impressed."

[7]He did not need calculus, since his formulation of the second law only concerned impulsive forces, rather than continually acting ones (Newton 1697, 1999: 113).

[8]This is one of the concepts that are unlikely to be represented *by* a CS, because it belongs to the meta-theory of such spaces.

[9]See Bramwell (2017) for a brief overview of dimensions in physics.

[10]Although analytic geometry is used to represent lengths, its symbols represent lengths implicitly divided by some reference length; this yields dimensionless scalar values, such that inhomogeneous terms can be added or equated (Schulman 2010).

In his introduction to DA, Sonin (1997) stresses the physical, empirical basis of comparison (similarity[11]), whereas other treatments tend to focus on an abstract, mathematical approach. Two clearly distinguishable objects, such as a book and a vase, can elicit the same response from a balance, establishing an equivalence in that respect. This emphasis on similarity suggests a deep connection between DA and CSs. Hence, it seems worthwhile to explore DA further and to take Sonin as our guide in the next subsections.

Though explicit dimensions are a theoretical product of modern physics, dimensioned quantities are commonly used in daily life (Borovik 2010). Even elementary school children use named numbers (numbers and parts of apples, cakes, *etc.*), which have a more sophisticated structure than, say, the rationals. Borovik (2010: 84) argues that *n* types of named numbers form a graded ring of Laurent polynomials: $\mathbb{Q}[x_1, x_1^{-1}, \ldots, x_n, x_n^{-1}]$, where each variable x_i stand for a particular name. Usually, only (some) monomials are interpreted as having real-life meaning. Further, Borovik (2010: 85) writes that "[P]hysicists love to work in the Laurent polynomial ring $\mathbb{R}[\text{length}^{\pm 1}, \text{time}^{\pm 1}, \text{mass}^{\pm 1}]$ because they love to measure all physical quantities in combinations (called 'dimensions') of the three basic units."[12,13]

Likewise, Tao (2012) regards physicists' DA as a parametric approach to representing dimensional quantities, since it relies on the choice of base units, as dimensional parameters. From the viewpoint of mathematics, Tao comments that physicists effectively consider all co-ordinations simultaneously by describing how their conventionally parametrized representation transforms under the relevant structure group encoding possible unit changes.[14]

8.2.2.1 Base Quantities and Units

Bridgman (1922) insisted on the conventional aspect of choosing units. As a result, Maxwell's notion of 'fundamental units' (see footnote 12) got replaced by 'base units' (Roche 1998). A physical property is considered to be a base quantity if (1) there is a physical operation to determine whether two objects or processes have an equal or an unequal quantity of this property and (2) there is an addition operation that defines the meaning of the sum of such quantities (Sonin 1997).[15] The latter

[11] See also Sterrett (forthcoming): "Similarity of systems is always relative to some kind of behaviour (e.g., kinetic, dynamic, electrical, thermal, magnetohydrodynamic)."

[12] Maxwell considered units of length, time, and mass as fundamental (Bramwell 2017).

[13] SI does not require fractional powers of basic units, but to represent unit systems that do (such as CGS), we need a richer mathematical structure.

[14] Applying Toa's view to SI, DA is the representation theory of the structure group {(M,L,t,T,I,n,li) | M,L,t,T,I,n,li $\in \mathbb{R}^+$}; see also Table 8.1.

[15] Since we use the colour CS as our guiding example, it is interesting that observe that Sonin (1997: 15) excludes colour as a physical quantity because the addition operation is not uniquely defined.

condition requires being able to construct (at least in principle) a third object or process of which the relevant property is expressed by the numerical sum of the quantities of the property present in two given objects or processes. The importance of the two operations is that they require manipulations in the physical world, rather than purely abstract calculations.[16]

These two operations allow us to choose a standard reference for a particular property: an object or process of which the relevant quantity is to be used as a unit of the quantity. Any quantity can then be expressed as "a multiple of a standard sample of its own kind, that is, to 'measure it in terms of a unit'" Sonin (1997: 15). "The measuring process consists of physically adding replicas of the unit and fractions thereof until the sum equals the quantity being measured" and counting the entire and fractional replicas. In this way, DA has become operationally defined.

We can now compare the definition of base quantities to Gärdenfors and Zenker's (2013: 1042) definition of a domain of the CS belonging to an empirical theory as a set of integral dimensions that are separable from all other dimensions. Their definition relies on the invariance relations of the relevant empirical theory, which are related to what scientists find sensible measurement operations (which magnitudes can be measured independent of others). Dimensions form a domain if their invariance class does not involve any additional dimensions.[17]

8.2.2.2 Physical Quantities, Derived Quantities and Dimensions

Some but not all formulas involving base quantities represent a physical quantity. Bridgman's (1922) "principle of absolute significance of relative magnitude" tells us which ones do.[18] First, any formula involving physical quantities can be rewritten as a relative formula, representing the ratio between the numerical values of the formula for two different objects or processes. If for a particular form of the formula, this ratio of any two objects or processes remains constant upon changes of the base unit sizes, then the formula corresponds with some physical quantity. In addition, it can be shown that to meet this condition it is both necessary and sufficient for a monomial formula to have the form of a power law: a real constant times real-valued powers of base quantities. However, not all quantities of this form need to be physically meaningful; *e.g.*, the 'square root of time' is a physical quantity according to this definition, although it cannot be represented physically by a particular object or process (Sonin 1997: 16–18).

[16]Besides physical operations that can be represented numerically, purely numerical operations can be defined, such that monomials of base quantities are defined.

[17]See also Sect. 8.2.3.1.

[18]Bridgman's principle is the rationale behind the aforementioned requirement that physical equations and terms therein should be homogenous. Recall that Galileo and Newton expressed equalities (verbally) in ratio form, which yields dimensionless quantities. Physically relevant dimensionless quantities are invariant under unit changes; hence, the importance of such quantities (like Reynolds number in fluid mechanics) in DA.

Besides base quantities, there is an "open-ended system of 'derived' physical quantities", that depend upon the base quantities via definitions or physical laws. Sonin (1997) distinguishes between two kinds of derived quantities:

(I) **Derived quantities of the first kind** are physical quantities that are not base quantities, but that can be expressed via a definition or a physical law in terms of base quantities or already established derived quantities. An example is kinetic energy, which depends on mass (base quantity) and velocity (derived quantity).

The term 'dimension' is introduced because of these derived quantities of the first kind, for which no physical units exist (Sect. 8.2.2.1), "but whose numerical values nevertheless depend on the choice of base units" (Sonin 1997: 18).

(II) **Derived quantities of the second kind** are base quantities and as such they could have served as fundamental units, but they were not selected by the unit system at hand. In addition, they are related by a definition or a general physical law to those that are. Examples in the SI include area and force.

Hence, both base quantities and derived quantities of the second kind depend on the choice of units. I propose the name baSIc quantities for base quantities that are not derived ones of the second kind under SI.

Concerning (II), both length and area fulfil the proposed definition of a base quantity. Of course, we are used to thinking of an area as a product of two lengths (or a two-dimensional integral), but Sonin (1997) stresses that direct physical comparisons only allow us to compare a given area to other areas. Given the definition, the existence of a numerical relation between areas and lengths does not establish that area is not a base quantity.[19] However, a unit system may be such that it includes a base unit for length but not one for area – as is indeed the case for units in the current SI.

The product definition for area is a numerical relation, rather than 'what area really means' (*cf.* Sonin 1997: 25).[20] In fact, physics does not deal with figuring out what any of these properties or qualities are in themselves, but only with (a) comparing them quantitatively (via the two operations for base quantities) and (b) with finding regularities in the relations between their quantities.[21]

[19]Euclidean geometry resulted from the observation of empirical regularities, which have been ubiquitous from the dawn of civilization that it is hard to recognize them as discoveries (Blåsjö 2016). It is conceivable that the physical world could have been such that the area of a rectangle does not (approximately) equal the product of its length and width, as, say, in a curved space where a non-Euclidean geometry applies.

[20]Tao (2012) has commented that, while it may be "geometrically natural to multiply two lengths to form an area" (but see footnote 19 for a comment on 'geometrically natural'), "it is less clear how to apply it for more general [dimensions]", such as the product appearing in the righthand-side of Einstein's formula $E = mc^2$.

[21]In this sense, all our empirical knowledge is comparative only and coherentism rather than foundationalism is the epistemology that fits naturalism best.

Table 8.1 gives an overview of the relations that exist between base and derived physical quantities, the choice of SI as the unit system, and the associated notion of dimensions.[22]

8.2.3 *From Dimensions to Spaces in Physics*

In the previous section, we reviewed the meaning of dimension in physics: it is an extension of the quality of base quantities (for which there exist physical objects are processes that can be used as units) to the quality of quantities that can be defined in terms of the former. Now, we investigate whether physical dimensions are like quality dimensions in a CS. In particular, we could combine all or some of them into a space and check whether it functions like a CS.

The seven basic dimensions of the SI[23] form a multiplicative group. The most straightforward suggestion is to use them as the basis of a space: representing them as orthogonal unit vectors, they span a seven-dimensional vector space.[24] If we take this to be the rational vector space, $\mathbb{Q}^7$, it can represent any SI dimension of the form $\mathsf{L}^{\alpha}\ \mathsf{M}^{\beta}\ \mathsf{T}^{\gamma}\ \mathsf{I}^{\delta}\ \Theta^{\varepsilon}\ \mathsf{N}^{\zeta}\ \mathsf{J}^{\eta}$ as a seven-dimensional vector $(\alpha, \beta, \gamma, \delta, \varepsilon, \zeta, \eta)$, where all the indices are rational numbers. The concept of force, for example, is represented by the vector (1,1,−2,0,0,0,0) in this space (see Fig. 8.2 and Moon and Spencer 1949: 497–498). Hence, this space can be thought of as a CS of dimensions of physical quantities in SI.[25]

---

[22]Sonin (1997: 32) points out that "only a few of the available universal laws are usually 'used up' to make base quantities into derived ones of the second kind." "There are many laws left with universal dimensional physical constants that could in principle be set equal to unity: the gravitational constant G, Planck's constant h, Boltzmann's constant k_B, the speed of light in vacuum c, etc." Although this leads to so-called natural units (*e.g.*, $c = 1$ as used in some branches of physics), the choice is far from unique.

[23]A revision of SI is due in 2018, such that the seven units will be defined in terms of seven constants that are defined rather than measured. Until the planned reform, the definition of the kilogram still depends on an artefact (the International Prototype of the Kilogram), as was previously the case for other units. Nothing in this exposition depends on the planned change. In this paragraph and in Table 8.1, the SI convention for the notation of dimensions is used (BIPM 2006, Sect. 8.2.3).

[24]Various suggestions of this kind can be found in the DA literature. I not aware a standard approach, but an early suggestion along these lines can be found in Moon and Spencer (1950): they considered dimensions as concepts and called their geometric approach 'idon theory' or 'idonics', but these names did not catch on. These authors have also developed a theory of 'holors' (mathematical quantities that are made up of one or more other independent quantities – a generalisation of tensors), which I want to flag here as potentially relevant for CSs.

[25]It should be noted that the number of baSIc dimensions (seven) is not a fundamental property of the physical world, but a matter of convention. It is possible to distinguish fewer basic dimensions; in fundamental physics, it is common to consider all quantities as scalar (dimensionless; see, *e.g.*, Duff et al. 2002: part III).

Table 8.1 Overview of relations between dimensioned physical quantities, SI as the unit system, and dimensions

Base quantities (with dimension symbol): Determined by physical operations		Defined quantities: Defined as monomials of base quantities			Non-homogeneous quantities	Undefined expressions
E.g., length (L), (gravitational) mass (M), duration (T), area (A), and (net) force (F)		General form of dimension: $a\,\mathsf{L}^{\alpha}\,\mathsf{M}^{\beta}\,\mathsf{T}^{\gamma}\,\mathsf{I}^{\delta}\,\Theta^{\varepsilon}\,\mathsf{N}^{\zeta}\,\mathsf{J}^{\eta}$, with $a \in \mathbb{R}\backslash\{0\}^{a}$ and $\alpha, \beta, \gamma, \delta, \varepsilon, \zeta, \eta \in \mathbb{Q}$ (at least one of which has to be non-zero)			*E.g.*, $a\,\mathsf{L}^{\alpha} + b\,\mathsf{M}^{\beta}$	*E.g.*, $\mathrm{Log}(\mathsf{L}^{\alpha})$ and $\mathrm{Sin}(\mathsf{M}^{\beta}\,[t]^{d})$
BaSIc quantities correspond to SI units	Derived quantities of the second kind	Derived quantities of the first kind			/	
Length (L), mass (M), duration (T), electric current (I), thermodynamic temperature (Θ), amount of substance (N), and luminous intensity (J)	*E.g.*, area (L^2), electric charge (IT), heat ($\mathsf{L}^2\mathsf{MT}^{-2}$), force ($\mathsf{LMT}^{-2}$)	*E.g.*, energy ($\mathsf{L}^2\mathsf{MT}^{-2}$), momentum ($\mathsf{LMT}^{-2}$)	Universal physical constants** *E.g.*, G, c, h	Local physical constants *E.g.*, g_{Paris}, k_{spring}		
Dimension equals the baSIc quantity*** Physical reference object or process given by SI convention	Dimension is the formulaic expression of dependence of the quantity upon baSIc quantities These may be without physical reference object or process, *e.g.*, $\mathsf{T}^{1/2}$				Not a dimension	

*To have a coherent set of quantities, a has to be unity for all derived quantities (as is the case for SI)
**The values of all physical constants "are fixed once the system of units is chosen" (Sonin 1997: 24). In addition, for universal constants, their "magnitudes are the same in all problems once the system of units is chosen" (Sonin 1997: 24)
***Observe that this is a special case of the corresponding sentence for derived quantities: the formula expresses identity

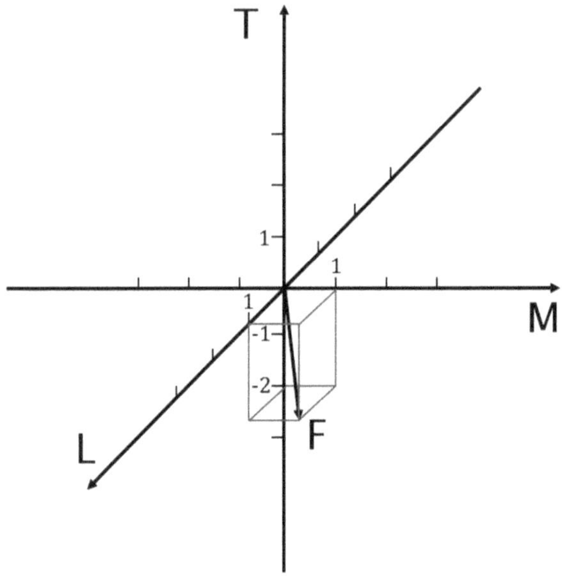

Fig. 8.2 The concept of force as a vector in the LMT-subspace of $\mathbb{Q}^7$

While using the seven baSIc dimensions as a basis is useful for doing DA, it is not the only viable way of organizing SI dimensions into a space. This raises the following questions:

(Q1) Should derived dimensions not be used as dimensions of a CS? Or is it precisely by using them that empirical regularities can be represented in a space in a particularly simple way?
(Q2) What do points in the space represent? What would be the concept that is represented by such a space? And do specific (convex) (hyper-)volumes of any of these spaces carry special meaning?
(Q3) DA as described so far looks particularly well-suited for classical physics, silently assuming Euclidean geometry throughout. What about special relativity (SR) and general relativity (GR)?

These questions will occupy us for most of the remaining chapter. We will address (Q1) in Sect. 8.2.3.1 and in Sect. 8.3, (Q2) in Sect. 8.2.3.1, and (Q3) in Sect. 8.4.4.

8.2.3.1 Invariance Under Unit Change Group

Regarding (Q1), Hulin (1980) compares two forms of DA: conventional DA (which relies on a unit system to establish relations for dimensional constants, usually SI) and tailor-made DA (optimized for the problem at hand, which cannot be done algorithmically). Hulin remarks that, in the latter case, the unit change group can be maximized, by eliminating spurious dimensional constants.

The invariance of physical relations under changes of units is actually a group invariance, which reflects that a unit system (in principle[26]) allows a free choice of base units and a dependent one for derived quantities. The composition law of unit changes is multiplicatory. By transforming physical quantities to their logarithms, the group of unit changes is transformed into an isomorphic continuous group of translations, with an additive composition law.

Moreover, Hulin (1980: 52) has given a geometrical interpretation of (conventional) DA: "[a] state of the studied system may be represented by a point P in an affine n-dimensional space $\mathcal{V}$". Hence, we could hypothesize this space to be the conventional CS of (classical) physics.[27] (We return to (Q1) in Sect. 8.3.)

On the other hand, and this point is related to (Q2), while the current SI only has seven basic units and associated dimensions, we might want to distinguish, *e.g.*, between lengths in different directions (L_x, L_y, L_z), as was proposed by Williams (1892) and Huntley (1952).[28] While this remains common practice in elementary applications of DA, it does complicate the interpretation of the space spanned by SI dimensions as a privileged CS. Huntley's suggestion was later refined by Siano (1985), who introduced orientational symbols that allow vector products. This enables rewriting the laws of physics as tensor equations (independent of the choice of a coordinate system).

Tao (2012) considers an alternative way (besides the coordinated one briefly described in Sect. 8.2.2) to represent dimensioned quantities, which doesn't require the choice of units as dimensional parameters. Instead, dimensioned objects "now live in an abstract mathematical space (*e.g.*, an abstract vector space), in which only a subset of the operations available to general-purpose number systems such as $\mathbb{R}$ or $\mathbb{R}^3$ are available, namely those operations which are 'dimensionally consistent' or invariant (or more precisely, equivariant) with respect to the action of the underlying structure group".

While these remarks on DA are suggestive of a unique, all-encompassing space of dimensions, this does not imply that all reasoning in classical physics is associated with a single CS. In the next section, we discuss examples of spaces that allow us to represent, *e.g.*, multiple positions of particles within a single system.

---

[26]Since unit systems are closely related to metrology, in practice considerations of accuracy, reliability, and reproducibility come into play as well.

[27]This is reminiscent of Raubal's (2004) formalization of psychological CSs as vector spaces but suggests opting for affine spaces instead. Affine spaces are generalizations of Euclidean spaces, which have no specific origin: they can be used to represent displacements (translations) or lengths, durations, *etc.* rather than absolute positions, points in time, *etc.*

[28]Moon and Spencer (1949) distinguished two basic length concepts (tangent and radial) to get rid of ambiguities (different concepts having the same dimensions if only one length dimension is considered).

8.2.3.2 Spaces from Classical Physics

So far, we have focused on dimensions used in physics. Now, we consider whether physics offers any readymade spaces to serve as CSs. We review three types of spaces readily used in classical physics, implicitly assuming units throughout:

1. **Coordinate space, L^3**

For a point particle, coordinate space is three-dimensional, represented by $\mathbb{R}^3$. The six-dimensional space $\mathbb{R}^3 \times \mathrm{SO}(3)$ can be used to represent the position and the orientation of a rigid body.

2. **Configuration space, L^{3N}**

To represent the configuration of N particles at a time t, we need N vectors in $\mathbb{R}^3$, or $3N$ coordinates: $\mathbf{r}_i(t) = (x_i(t), y_i(t), z_i(t))$. Usually in physics, there are constraints on the movement of particles in a system. If there are C independent constraints, there are $n = 3N - C$ degrees of freedom and n generalized coordinates: $\mathbf{q}(t) = (q_1(t), \ldots, q_n(t))$. By graphing the energy function across the configuration space, we obtain the energy landscape: a hypersurface in $\mathbb{R}^{3N+1}$. Reasoning by analogy to an actual landscape in a gravitational potential enables us to interpret critical points easily (*e.g.*, minima are stable).

3. **Phase space, $\mathsf{L}^{3N} \times (\mathsf{LMT}^{-1})^{3N}$**

For N point particles, specifying position and momentum of each requires $6N$ scalars. Microstates of a system correspond to points; trajectories (*e.g.*, orbits) parametrized by time correspond to curves; and ensembles correspond to regions.

Of these three contenders, phase space is the most promising candidate to be considered as a CS, as we will argue in the next section.

8.2.3.3 Phase Spaces

As an alternative approach to finding CSs in physics, we can try to generalize from the example that we have already identified in Sect. 8.2.1. The phase diagrams from thermodynamics can be generalized to phase spaces: each point of such a space represents a possible state in which a given system can exist.

In statistical mechanics, a system of N gas particles in a three-dimensional space can be described in a $6N$-dimensional phase space. In realistic examples, N is of the order of a mole, so the number of dimensions of the phase space is huge. Statistical physics studies how coarse-graining a particle-based phase space can be related to an effective phase diagram of macroscopically accessible, thermodynamic variables, leading to a massive reduction of dimensions.

Outside of mechanics and statistical physics, phase spaces can represent other dimensions.[29] They are used to represent the temporal evolution (dynamics) of systems. One can represent the qualitative evolution (change in time dimension) of objects using trajectories in a phase space. Notice that this usage extends the concept of trajectories (including orbits) from real space to phase space. Still, the concept of 'orbit' doesn't 'live' in the phase space of a particular system (recall Sect. 8.2.1). Conceptual similarity to, *e.g.*, a prototypical eccentric orbit is assessed across phase spaces, which suggests the relevant CS to be a different space. Similar remarks apply to the concept of critical points.

The work of Masterton et al. (2017) is suggestive of the following identity thesis: the phase spaces of various branches of physics *are* CSs. To evaluate the merits of the identity thesis, let us consider both corroborating and opposing evidence in the next four paragraphs.

Sufficiency? Masterton et al. (2017) state that addition of constraints on allowed trajectories on a CS leads to a theory of the systems in that space. Hence, different theories may share the same CS, used to predict (or retrodict) future (or past) qualities from a given state. However, this may be misleading (or at least not fully general) since the phase space co-depends on the system (including, *e.g.*, the number of particles N). This suggests considering an implicit meta-space, in which the properties of the system (including N) itself are varied.

Necessity? One reason to consider phase spaces as candidate CSs is that they are a generalization of phase diagrams. We might look at a different generalization of phase diagrams: are there other types of diagrams used in physics that can be interpreted as CSs? Let us just consider one example: the Hertzsprung-Russell diagram (Fig. 8.3). In a pre-theoretic context, apparent positions of stars are their most salient properties, possibly followed by brightness and colour. The Hertzsprung-Russell diagram promotes the salience of the latter two dimensions and immediately visualizes a connection between them, which led to an important insight into stellar evolution. This example is quite similar to the recategorization brought about by Linnaeus's work in plant biology, as discussed by Gärdenfors (2000, 2004: 2018–219). Although the Hertzsprung-Russell diagram serves a purpose that theoretic CSs are required to serve, it is usually not regarded as a phase space. This suggests that being a phase space may not be necessary for being a CS and that examining other diagrams from the physical sciences might yield many additional candidate CSs.

Vacuity? Masterton et al. (2017) remind us that, in the tradition of Boltzmann, phase spaces have been used to represent degrees of freedom of a complex system and to model chaotic behaviour. Yet, history of science shows that theories are not originally presented in their phase space formulation; it is part of the development of a theory to find such a formulation. Classical mechanics (CM), SR theory, and quantum mechanics (QM) are all modelled using phase spaces. Masterton et al.

[29]In the case of discrete dimensions, the corresponding phase space is called a 'state space'. A two-dimensional phase space is called a 'phase portrait'.

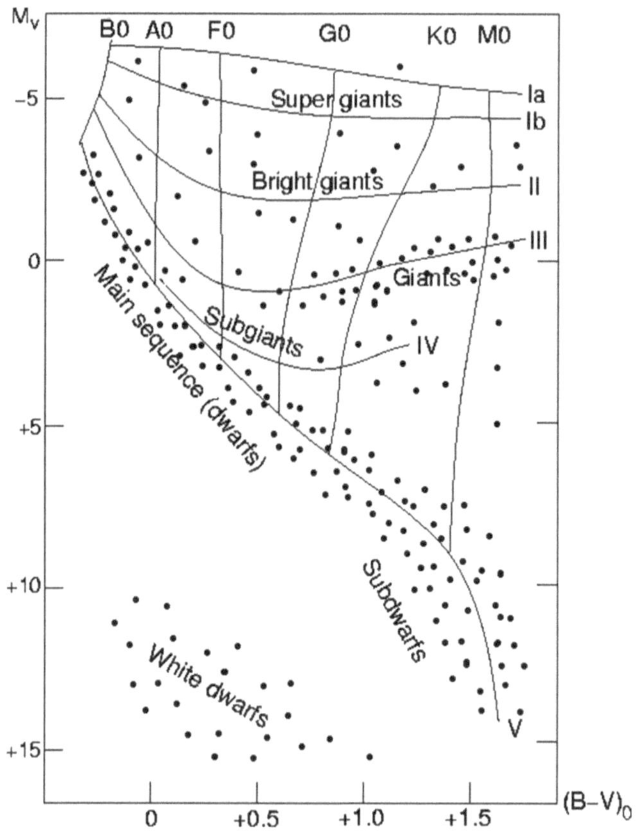

Fig. 8.3 The Hertzsprung-Russell diagram represents observed stars in terms of their B–V colour index (horizontal axis, related to effective temperature) and absolute magnitude or brightness (vertical axis), showing clusters and sequences. (Image by Wikipedia user Looxix)

(2017) observe that in the case of QM, the phase space formulation is not very practical and that for GR theory, no formulation has been found (yet). It takes time before the phase space of a theory is developed. So, what about the CS of the theory in the meantime: doesn't the theory have a CS, or does another space function as its CS *ad interim*? (*Cf.* previous discussion point.)

Convexity? We mentioned that criterion P requires natural psychological categories to correspond with convex regions in CSs. It is usually not assumed that the criterion applies to theoretical CSs, for one, because physicists allow coordinate transformations. Yet, if we do look for it in phase spaces, we arrive at ensembles, which are often defined at a given time where they belong to a convex region. The systems belonging to an ensemble typically do not remain in a convex region: their temporal evolution is studied to determine chaoticity.

8.3 Examining the Dimension of Force in Newtonian Particle Physics

To answer (Q1) more fully, we look at the physical domain of force. It has already been debated in the literature on CSs whether Newtonian force is a fundamental dimension or not. Here, we briefly revise the discussion and then add the perspective from DA.

Gärdenfors and Zenker (2013) attribute four separable domains to Newtonian particle mechanics: space, time, mass, and force, which correspond to, respectively: three Euclidean dimension, one dimension isomorphic to the real number line, one dimension isomorphic to the positive real number half-line, and a three-dimensional Euclidean vector space;[30] that is eight dimensions in total.[31] Because they assume that forces cannot be determined without applying the theory itself, Gärdenfors and Zenker (2013: 1049) consider it debatable whether the domain of force belongs to the CS. Earlier, Gärdenfors (2000: 216) had observed that Newton's law connects all the domains (on the view that the relevant CS is eight-dimensional), or defines force (on the view that the relevant CS is five-dimensional).

Using the viewpoint of DA, we can shed some light on this discussion. In Table 8.1, force is classified as a derived quantity of the second kind: a base quantity, yet not independent from dimensions that receive a base unit in the SI convention.[32] Since the latter part is conventional, the five-dimensional CS should be considered as partially conventional, too. In as far as this is undesirable, the eight-dimensional CS is to be preferred. As Gärdenfors (2000: 216) observed, Newton's second law defines a hypersurface in the eight-dimensional CS: this is the empirical content of the theory. These two points suggest leaving in dimensions corresponding to derived quantities of the second kind, thereby answering (Q1).

On the other hand, we would like to raise an additional complication. Although Newton's original formulation of the second law indeed only requires one vector (*cf.* footnote 7), in general, we need a force *field*, which specifies a force vector (three degrees of freedom) at each point of three-dimensional space. Gärdenfors and Zenker (2013, Sect. 8.4.3) discuss scalar and vector fields in the context of electromagnetism. They mention that vector fields can be identified with functions going from a three-dimensional space (S) to the set containing the values assigned to each point (three-dimensional F). They only mention the sum of dimensions of the domain and the co-domain (six for vector fields), not the dimension of these function spaces. Yet, it seems to me that it is the function space that should be regarded as the

[30] If we read this assuming implicit SI units, there are tacit ratios involved to arrive at real values (*cf.* footnotes 10 and 18).

[31] See also Gärdenfors 2000: 215–219.

[32] The extant literature on CSs also considered velocity, acceleration, momentum, energy, and work as derived quantities, claiming that to define them, we need to combine magnitudes belonging to different domains. Again, DA adds a helpful distinction between merely defined quantities and derived quantities of the second kind.

relevant CS. In set theory, it is common to represent the set of all functions from S to F as F^S and the (algebraic) dimension of a set is the cardinality of a set of algebraic basis vectors. Considering that the set of all functions from $\mathbb{R}$ to $\mathbb{R}$ ($\mathbb{R}^{\mathbb{R}}$) already has an infinite algebraic basis and hence an infinite algebraic dimension, it should be clear that the same holds for F^S.

8.4 The Geometrization of Time and Its Discontents

To address (Q3), we will focus on the physical dimension of time.[33]

8.4.1 *Newtonian Mechanics*

Absolute space and time are sometimes called the 'Newtonian theatre', since in the Newtonian worldview events take place within space and time without constituting or affecting them. We already observed in Sect. 8.3 that Gärdenfors and Zenker (2013) reconstruct the CS as a five- or eight-dimensional space, with a one-dimensional temporal domain and a three-dimensional spatial domain.

8.4.2 *Reformulations of CM*

We can represent the change of position of a point mass in at least two ways: by parametrizing its trajectory by a time coordinate (within a three-dimensional representation) or by representing time as a fourth, separable Euclidean dimension.

Although Newton presupposed the absoluteness of space and time, it later became clear that Newton's laws do not require this. The laws are invariant under coordinate transformations that amount to switching to another inertial frame: these transformations form the Galilean group. Hence, a relative notion of space and time suffices for classical particle physics.

After SR had received its four-dimensional formalization by Minkowski (see Sect. 8.4.4.1), Cartan developed a neo-Newtonian theory, reconstructing Newtonian theory in a four-dimensional spacetime. On Cartan's proposal, distance relation between points in different hyperplanes don't exist, only among points in each hyperplane. Hence, there are no transtemporal spatial relations and there are no absolute velocities. Newton-Cartan spacetime is structurally similar to the spacetime

[33]We should not expect to reveal time's true nature by studying it as a physical dimension. As Sonin (1997:14) reminds us: "What time 'is' has no relevance in the present context, only the defining operations matter."

of GR: it also has curvature, but, unlike in SR, there is no maximum velocity and there is absolute simultaneity.

Gärdenfors and Zenker (2013: 1050) remark that Newtonian physics is also empirically equivalent to Hamiltonian CM. Rather than a force (field), it requires a momentum vector (in $\mathbb{R}^3$) as one of the initial conditions. Moreover, the total energy (or 'Hamiltonian') of the system is conceptually much more central in the latter: a change in the importance of a domain.

The domains of Hamiltonian mechanics are time t, positions **q**, and momenta **p**. The configuration space of a system is a manifold M that represents positions **q** in the physical space. The momentum **p** is an element of the cotangent space $T_{\mathbf{q}}{}^{*}M$. In the case of Hamiltonian mechanics, the Lie algebra of the phase space transformation operations is the Lie algebra of the Galilean group.

A system of *N* point masses moving in three-dimensional physical space can be represented as a single point moving in a 6*N*-dimensional phase space, where the temporal evolution of the system is described by the Hamilton equations. Typically, time is used to parametrize trajectories in phase space, as (**q**(t), **p**(t)): this can again be regarded as a projection of an extra dimension.

8.4.3 Representing Time as a Dimension in Phase Spaces

In Sect. 8.2.3.3, we considered whether phase spaces are CSs. In any case, one can represent the qualitative evolution of objects or systems using trajectories in both. Alternatively, time can be represented as an additional dimension, as we just saw for CM.

This practice is utterly familiar, also outside science: in comics, chronology is represented spatially by subsequent frames (McCloud 1993, Chap. 4). Cultural entrenchment makes it hard to notice, let alone challenge it. Yet, some authors have objected against viewing time as merely a geometrical dimension, including continental philosopher Henri Bergson and contemporary philosopher of physics Tim Maudlin (2007,712: 107), who believes: "that it is a fundamental, irreducible fact about the spatio-temporal structure of the world that time passes. [...] The passage of time is an intrinsic asymmetry in the temporal structure of the world, an asymmetry that has no spatial counterpart."

8.4.4 Time in Theories Beyond Classical Physics

8.4.4.1 Special Relativity Theory (SR)

Gärdenfors and Zenker (2013: Sects. 8.4.2 and 8.4.3) provided a detailed account of the conceptual changes constituting intermediate steps from Newtonian physics to SR. Based on the Lorentz group of transformations, Masterton et al. (2017) conclude that the phase space of SR is the eight-dimensional cotangent space (*cf.* Amelino-Camelia et al. (2011) discussed below).

Tao (2012) generalized the structure group mentioned in Sect. 8.2.2 to include vector-quantities (rather than just scalars) as dimensional units and to spacetime. This allows him to generalize DA to include SR: in this case the relevant structure group is the group of Minkowski spacetime isometries (a semidirect product of the Lorentz group and the translations) and "the principle of SR can be interpreted as the assertion that all physical quantities transform cleanly with respect to this group action (as opposed to having a hybrid transformation law, or one which privileges a specific reference frame)".

8.4.4.2 What About GR and Quantum Gravity?

As Masterton et al. (2017) remind us, there is no phase space formulation of GR (yet). We merely have the phase space of SR as a local approximation to it. Tao (2012) put it as follows: "if the [GR] spacetime is topologically non-trivial, one needs to work with local coordinate charts (or atlases of such charts, and the structure group needs to be relaxed to a looser concept, such as that of a pseudogroup) rather than a single global coordinate chart, which greatly complicates the parametrisation space".

A phase space formulation of GR is believed to be an important step towards a theory that unifies or extends GR and QM. Amelino-Camelia et al. (2011) proposed to use an eight-dimensional phase space, as the basis of such a theory for quantum gravity. These are to be considered as one domain, which allows to represent entanglement between spacetime and momentum space.

In terms of salience of conceptual dimensions, it is interesting that they write the following: "Our most fundamental measurements are the energies and angles of the quanta we emit or absorb, and the times of those events. Judging by what we observe, we live in energy-momentum space", *i.e.*, phase space, and they "take the point of view that spacetime is an auxiliary concept which emerges when we seek to define dynamics in momentum space."

There are proposals for fundamental theories in which space or spacetime is emergent. Especially the latter is very difficult to imagine: we already mentioned Bergson and Maudlin, who take change and the passage of time to be irreducible aspects of the world. In addition, if spacetime is emergent, this sense of emergence should not be thought of in a temporal way. As Sklar (1983: 45) expressed it: "What could possibly constitute a more essential, a more ineliminable, component of our conceptual framework than that ordering of phenomena which places them in space and time? The spatiality and temporality of things is, we feel, the very condition of their existing at all and having other, less primordial, features. A world devoid of color, smell or taste we could, perhaps, imagine. [...] But a world not in time? A world not spatial? Except to some Platonists, I suppose, such a world seems devoid of real being altogether."

Yet, there currently are proposals that aim to do just this. One of them is 'loop quantum cosmology', in which spatiotemporal relations are regarded as approximate, effective properties arising from underlying spin networks.[34] The general idea common to such proposals is that there is at least one abstract dimension (not corresponding to anything we can measure), which gives rise to effective structures associated with less abstract, empirically accessible degrees of freedom.

8.5 Final Remarks

One of the goals of science is to identify hypotheses that perform best at accurately and succinctly describing our current evidence. In order to do this, it tries to distinguish more fundamental regularities and natural kinds from merely accidental categorizations and gerrymandered groups. Whereas DA helps to tell apart more natural dimensions (that allow for suitable invariances) from less natural ones, CSs may be a good way to investigate science's ongoing search for natural kinds and associated concepts.

If we are interested in studying changes associated with concepts in the history of physics, in order to establish the interplay between continuity and more abrupt changes, identifying CSs may be a worthwhile approach (Masterton et al. 2017). However, we should keep in mind that some central concepts of physics (such as temperature) are most naturally represented by dimensions of spaces (such as phase spaces), rather than as particular regions thereof. (We discussed a space of physical dimensions, consistent with the current SI, in Sect. 8.2.3.)

In addition, at a given time, physicists use some concepts across various abstract spaces, without representing those concepts in an abstract space. As we pointed out in Sect. 8.2.1, such concepts don't 'live' on the spaces to which they are associated, but they are crucial for developing analogies and mappings between domains in physics. It may be worthwhile to identify and track such concepts across CSs. It might even be possible to make explicit the spaces in which these concepts 'live', even though the physicist is unaware of them (like the example of 'orbit' Sect. 8.2.3.3), much like most people are unaware of the organization of their conceptual colour space. For instance, many physicists have no problem applying the concept of an orbit or, more generally a trajectory, transferring it from coordinate space to more abstract spaces (most notably, phase spaces), but these are not the CSs on which the concepts orbit and trajectory 'live'. Likewise, physicists may have some vague idea of a space of all possible physical theories, without being able to (or needing to) specify its dimensions. This is connected to the problem of new theories in Bayesian confirmation theory, and in general thinking about such a space is a meta-scientific project (Wenmackers and Romeijn 2016).

---

[34]Huggett and Wüthrich (2013) provide an introduction suitable for philosophers of science.

Finally, remark that, once a new CS emerges, there can be a tendency for reconstructing or embedding the old meaning in the new CS.[35] Such post hoc reconstructions may make the old position look silly and/or downplay the radicalness of the shift. The progression of from space, over spacetime, towards energy-momentum-spacetime (Sect. 8.4.4.2) suggests that there might be something akin to a gas law for concepts: a scientific concept tries to escape to as high-dimensional an integral domain as possible. Once a concept has been integrated in such a domain, it is nearly impossible to contain it again in its old corner of a conceptual Flatland. Likewise, the meaning of phase, space, and dimension have all been radically altered from their origins to include ever more metaphorical and abstract meanings.[36] As Bramwell (2017) put it, "our changing concept of dimensions reflects the evolution of physics itself". These do not seem appropriate concepts for studying them within CSs, but it is interesting to observe how they co-evolved together with those concepts of which the evolution may be studied through reconstructions of the involved CSs.

Acknowledgments I am grateful to the organizers and participants of the workshop "Conceptual Spaces at Work" (2016) for giving me the opportunity to present and discuss my early ideas on this topic. I also thank Joel Parthemore, Danny Vanpoucke, and an anonymous referee for their helpful comments on an earlier version of this text. Part of this project was supported by a grant from the FWO (Research Foundation – Flanders) through grant number G0B8616N.

References

Amelino-Camelia, G., Freidel, L., Kowalski-Glikman, J., & Smolin, L. (2011). Principle of relative locality. *Physical Review D, 84*, 084010.

Blåsjö, V. (2016). The how and why of constructions in classical geometry. *Nieuw Archief voor Wiskunde, 5*(17), 283–291.

Borovik, A. (2010). *Mathematics under the microscope*. London: American Mathematical Society.

Bramwell, S. T. (2017). The invention of dimension. *Nature Physics, 13*, 820.

Bridgman, P. W. (1922). *Dimensional analysis*. New Haven: Yale University Press.

Bureau International des Poids et Mesures. (2006). *The international system of units (SI)* (8th ed.).

Carey, S., & Spelke, E. (1994). Domain-specific knowledge and conceptual change. In L. A. Hirschfeld & S. A. Gelman (Eds.), *Mapping the mind* (pp. 169–200). Cambridge: Cambridge University Press.

[35]For example, we can present Euclidean geometry as a special case of curved geometry, with the curvature equal to zero. However, this is not a suitable reconstruction of the conceptual change involved in developing the first non-Euclidean theory, which required recognizing spatial curvature as a valid concept in the first place. And once curvature has been accepted as a variable, it is equally hard to imagine reasoning without it or to appreciate how counterintuitive its introduction must have been starting out from a strictly Euclidean viewpoint.

[36]See de Courtenay (2015) for an account of how Fourier's treatment of dimensional analysis and the standardisation of physical units have profoundly changed the meaning of mathematical equations in physics.

de Courtenay, N. (2015). The double interpretation of the equations of physics and the quest for common meanings. In O. Schlaudt & L. Huber (Eds.), *Standardization in measurement* (pp. 53–68). London: Pickering & Chatto.

Douven, I., Wenmackers, S., Jraissati, Y., & Decock, L. (2017). Measuring graded membership: The case of color. *Cognitive Science, 41*, 686–722.

Duff, M. J., Okun, L. B., & Veneziano, G. (2002). Trialogue on the number of fundamental constants. *Journal of High Energy Physics, 03*, 023.

Fourier, J. (1822). *Théorie analytique de la chaleur*. Paris: Didot.

Gärdenfors, P. (2000, 2004). *Conceptual spaces: The geometry of thought* (2nd ed.). Cambridge, MA: MIT Press.

Gärdenfors, P., & Zenker, F. (2013). Theory change as dimensional change: Conceptual spaces applied to the dynamics of empirical theories. *Synthese, 190*, 1039–1058.

Huggett, N., & Wüthrich, C. (2013). Emergent spacetime and empirical (in)coherence. *Studies in History and Philosophy of Modern Physics, 44*, 276–285.

Hulin, M. (1980). Dimensional analysis: some suggestions for the modification and generalisation of its use in physics teaching. *European Journal of Physics, 1*, 44–55.

Huntley, H. E. (1952). *Dimensional analysis*. London: MacDonald & Co.

Macagno, E. O. (1971). Historico-critical review of dimensional analysis. *Journal of the Franklin Institute, 292*, 391–402.

Masterton, G., Zenker, F., & Gärdenfors, P. (2017). Using conceptual spaces to exhibit conceptual continuity through scientific theory change. *European Journal for Philosophy of Science, 7*, 127–150.

Maudlin, T. (2007). *The metaphysics within physics*. Oxford: Oxford University Press.

McCloud, S. (1993). *Understanding comics*. Northampton: Tundra Publishing.

Moon, P., & Spencer, D. E. (1949). A modern approach to 'dimensions'. *Journal of the Franklin Institute, 248*, 495–521.

Moon, P., & Spencer, D. E. (1950). A geometric treatment of 'dimensions' in physics. *Canadian Journal of Research, 28*, 268–280.

Newton, I. (1697, 1999). *Principia Mathematica*. (I. B. Cohen & A. Whitman, Trans.). University of California Press.

Raubal, M. (2004). Formalizing conceptual spaces. In *Formal ontology in information systems, proceedings of the third international conference (FOIS 2004)* (Vol. 114, pp. 153–164).

Roche, J. J. (1998). *The mathematics of measurement: A critical history*. New York: Athlone.

Schulman, J. (2010). *A history of units and dimensional analysis*. Unpublished manuscript. Retrieved February 25, 2019, http://joschu.net/docs/history-of-units.pdf.

Siano, D.B. (1985). "Orientational analysis–a supplement to dimensional analysis–I" and "Orientational analysis, tensor analysis and the group properties of the SI supplementary units–II". *Journal of the Franklin Institute, 320*, 267–283 and 285–302.

Sklar, L. (1983). Prospects for a causal theory of space-time. In R. Swinburne (Ed.), *Space, time, and causality* (Vol. 157, pp. 45–62). Synthese Library.

Sonin, A. A. (1997 1st ed., 2001). *The physical basis of dimensional analysis* (2nd ed). Cambridge, MA: MIT Press.

Sterrett, S. G. (forthcoming). Theory of dimensions. In E. Knox & A. Wilson (Eds.), *Routledge companion to the philosophy of physics*. Routledge.

Tao, T. (2012). Retrieved February 25, 2019, https://terrytao.wordpress.com/2012/12/29/a-mathematical-formalisation-of-dimensional-analysis/.

Wenmackers, S., & Romeijn, J.-W. (2016). New theory for old evidence. *Synthese, 193*, 1225–1250.

Williams, W. (1892). On the relation of the dimensions of physical quantities to directions in space. *Philosophical Magazine, 5*(34), 234–271.

Chapter 9
Interacting Conceptual Spaces I: Grammatical Composition of Concepts

Joe Bolt, Bob Coecke, Fabrizio Genovese, Martha Lewis, Dan Marsden, and Robin Piedeleu

Abstract The categorical compositional approach to meaning has been successfully applied in natural language processing, outperforming other models in mainstream empirical language processing tasks. We show how this approach can be generalized to conceptual space models of cognition. In order to do this, first we introduce the category of convex relations as a new setting for categorical compositional semantics, emphasizing the convex structure important to conceptual space applications. We then show how to construct conceptual spaces for various types such as nouns, adjectives and verbs. Finally we show by means of examples how concepts can be systematically combined to establish the meanings of composite phrases from the meanings of their constituent parts. This provides the mathematical underpinnings of a new compositional approach to cognition.

9.1 Introduction

How should we represent concepts and how can they be composed to form new concepts, phrases and sentences? These questions are fundamental to cognitive science and thereby human-like artificial intelligence. *Conceptual spaces theory* gives a way of describing structured concepts (Gärdenfors 2004, 2014), not

This paper is a significantly extended version of the workshop paper Bolt et al. (2016).
Authors in alphabetical order.

J. Bolt · B. Coecke · F. Genovese · D. Marsden · R. Piedeleu
Quantum Group, Department of Computer Science, University of Oxford, Oxford, UK
e-mail: josef.bolt@cs.ox.ac.uk; coecke@cs.ox.ac.uk; fabrizio.genovese@cs.ox.ac.uk; daniel.marsden@cs.ox.ac.uk; robin.piedeleu@cs.ox.ac.uk

M. Lewis (✉)
Institute for Logic, Language, and Computation, University of Amsterdam, Amsterdam, The Netherlands
e-mail: m.a.f.lewis@uva.nl

M. Kaipainen et al. (eds.), *Conceptual Spaces: Elaborations and Applications*, Synthese Library 405, https://doi.org/10.1007/978-3-030-12800-5_9

starting from linguistic assumptions, but from cognitive considerations about human reasoning. Conceptual spaces describe a "semantics of the mind", modelling mental descriptions of concepts. The key idea is that human beings represent concepts geometrically in certain fundamental domains of understanding such as space, motion, taste and colour. These domains are combined to form a *conceptual space* describing the features of interest. A concept is then described by convex subsets of the relevant domains. The convexity requirement can be seen as a means of identifying robust, meaningful concepts. For example, if two points in some space are considered to represent the colour red, then intuitively we would expect that every point "in between" would also be considered red. Conceptual spaces provide a middle ground between symbolic and connectionist representations of concepts, oriented towards tasks of interest in cognitive science, such as categorization and assessing similarity.

Categorical compositional distributional models (Coecke et al. 2010) successfully exploit the compositional structure of natural language in a principled manner, and have outperformed other approaches in Natural Language Processing (NLP) (Grefenstette and Sadrzadeh 2011; Kartsaklis and Sadrzadeh 2013). The approach works as follows. A mathematical formalization of grammar is chosen, for example *Lambek's pregroup grammars* (Lambek 1999), although the approach is equally effective with other categorial grammars (Coecke et al. 2013a). Such a categorial grammar allows one to verify whether a phrase or a sentence is grammatically well-formed by means of a computation that establishes the overall grammatical type, referred to as *a type reduction*. The meanings of *individual* words are established using a distributional model of language, where they are described as vectors of co-occurrence statistics derived automatically from corpus data (Lund and Burgess 1996). The categorical compositional distributional programme unifies these two aspects of language in a compositional model where grammar mediates composition of meanings. This allows us to derive the meaning of sentences from their grammatical structure, and the meanings of their constituent words. A concrete example of this is as follows. An intransitive verb such as *kick* is represented as a linear map – i.e., a matrix – from the noun space to the sentence space, and a noun such as *horses* as a vector in the noun space. The meaning of the intransitive sentence *horses kick* is then computed as matrix-vector multiplication of a vector representing *horses* and a matrix representing *kick*. The key insight that allows this approach to succeed is that both pregroups and the category of vector spaces carry the same abstract structure (Coecke et al. 2010), and the same holds for other categorial grammars since they typically have a weaker categorical structure.

The abstract framework of the categorical compositional scheme we discuss here is broader in scope than just NLP. It can be applied in other settings in which we wish to compose meanings in a principled manner, guided by structure. The outline of the general programme is as follows:

1. (a) Choose a compositional structure, such as a pregroup or any other categorial grammar.
   (b) Interpret this structure as a category, the **grammar category**.

2. (a) Choose or craft appropriate meaning or concept spaces, such as spaces of propositions, vector spaces, density matrices (Bankova et al. 2017; Piedeleu et al. 2015) or conceptual spaces.
   (b) Organize these spaces into a category, the **semantics category**, with the same abstract structure as the grammar category.
3. Interpret the compositional structure of the grammar category in the semantics category via a functor preserving the type reduction structure.
4. Bingo! This functor maps type reductions in the grammar category onto algorithms for composing meanings in the semantics category.

In order to move away from vector spaces, we construct a new categorical setting for interpreting meanings which respects the important convex structure emphasized in conceptual spaces theory. We show that this category has the necessary abstract structure required by categorical compositional models. We then construct convex spaces for interpreting the types for nouns, adjective and verbs, where a convex space is one in which it is possible to form convex combinations of elements. Finally, this allows us to use the reductions of the pregroup grammar to compose meanings in conceptual spaces. We illustrate our approach with concrete examples, and go on to discuss directions for further research.

9.2 Categorical Compositional Meaning

In this section we describe the details of the categorical compositional approach to meaning. We provide examples of semantic categories and grammar categories, and show the general method by which the grammar category induces a notion of concept composition in the semantic category.

9.2.1 Monoidal Categories

We begin by briefly reviewing some of the category theory underpinning categorical compositional models. This definition becomes a little involved, however, monoidal categories can be interpreted in an intuitive manner and represented using a purely diagrammatic calculus.

Definition 1 A **monoidal category** is a tuple $(\mathbf{C}, \otimes, I)$ where

- $\mathbf{C}$ is a category, meaning that:
  - $\mathbf{C}$ has a collection of **objects** $A, B, \ldots$ and each ordered pair of objects (A, B) has a collection of **morphisms** $f : A \to B$
  - For each triple of objects (A, B, C) and morphisms $f : A \to B$, $g : B \to C$ there is a **sequential composite** $g \circ f : A \to C$ that is associative, i.e.

$$h \circ (g \circ f) = (h \circ g) \circ f$$

- for each object A there is an **identity morphism** $1_A : A \to A$ such that for $f : A \to B$

$$f \circ 1_A = f \quad and \quad 1_B \circ f = f$$

- for each ordered pair of objects (A, B), there is a **composite object** $A \otimes B$, and we moreover require that:

$$A \otimes (B \otimes C) \cong (A \otimes B) \otimes C$$

where $\cong$ means 'is isomorphic to'.
- there is a **unit object** I, which satisfies

$$I \otimes A \cong A \cong A \otimes I$$

- for each ordered pair of morphisms $f : A \to B$, $g : B \to C$ there is a **parallel composite** $f \otimes g : A \otimes B \to C \otimes D$ which satisfies:

$$(g_1 \otimes g_2) \circ (f_1 \otimes f_2) = (g_1 \circ f_1) \otimes (g_2 \circ f_2) \tag{9.1}$$

For a precise statement and discussion of the above definition, we direct the reader to Mac Lane (1971). A more gentle introduction can be found in Coecke and Paquette (2011).

For the purpose of our paper, the objects of a monoidal category should be thought of as *system types*. A morphism $f : A \to B$ is then a process taking inputs of type A and giving outputs of type B. The object $A \otimes B$ represents the systems A and B composed in *parallel*. Hence, a morphism $f \otimes g : A \otimes B \to C \otimes D$ is to be thought of as running the process $f : A \to C$ *whilst* running the process $g : B \to D$. The object I is thought of as the trivial system. To borrow the cooking example given in Coecke and Paquette (2011), the sequential composite $f_2 \circ f_1$ can be thought of as running one process and then another: boiling potatoes and then mashing them, whereas the tensor product $f_1 \otimes g_1$ can be thought of as performing two processes in parallel: boiling potatoes whilst slicing steak. Then, the sequential composite of two parallel composite morphisms, $(g_1 \otimes g_2) \circ (f_1 \otimes f_2)$ can be thought of as boiling potatoes whilst slicing steak, and then mashing potatoes whilst frying steak. Equation 9.1 says that this is the same as boiling and then mashing potatoes whilst slicing and then frying steak.

Example 1 The category **Rel** of sets and relations is monoidal. The tensor $\otimes$ is the Cartesian product and I is any singleton set $\{\star\}$.

Example 2 The category $\mathbf{FdVect}_{\mathbb{R}}$ of finite dimensional real vector spaces and linear maps is monoidal. The tensor $\otimes$ is the tensor product, the trivial system I is the one-dimensional real vector space $\mathbb{R}$.

Monoidal categories admit an elegant and powerful graphical notation that we will exploit heavily in later sections. In this notation, an object A is denoted by a wire:

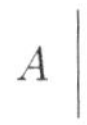

A morphism $f : A \to B$ is represented by a box:

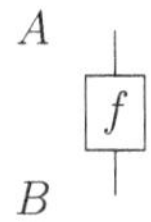

If $g : B \to C$, the composite $g \circ f : A \to C$ is given by:

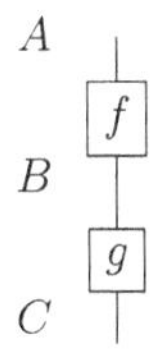

If $h : A \to B$ and $k : C \to D$, the morphism $h \otimes k : A \otimes C \to B \otimes D$ is depicted by:

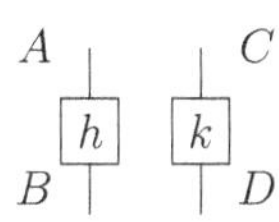

The trivial system I is the empty diagram. Morphisms $u : I \to A$ and $v : A \to I$ are drawn respectively as

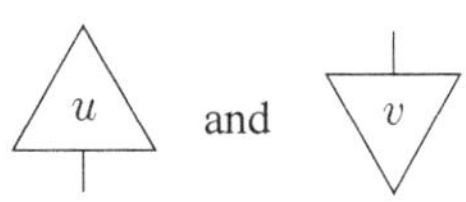

These special morphisms are referred to as **states** and **effects** respectively.

9.2.2 Compact Closed Categories

A specific class of monoidal categories, the compact closed categories, will be of particular importance.

Definition 2 A monoidal category $(C, \otimes, I)$ is **compact closed** if for each object $A \in C$ there are objects $A^l, A^r \in C$ (the **left** and **right duals** of A) and morphisms

$$\eta^l_A : I \to A \otimes A^l \qquad \eta^r_A : I \to A^r \otimes A$$

$$\epsilon^l_A : A^l \otimes A \to I \qquad \epsilon^r_A : A \otimes A^r \to I$$

satisfying the snake equations

$$(1_A \otimes \epsilon^l_A) \circ (\eta^l_A \otimes 1_A) = 1_A \qquad (\epsilon^r_A \otimes 1_A) \circ (1_A \otimes \eta^r_A) = 1_A$$

$$(\epsilon^l_A \otimes 1_{A^l}) \circ (1_{A^l} \otimes \eta^l_A) = 1_{A^l} \qquad (1_{A^r} \otimes \epsilon^r_A) \circ (\eta^r_A \otimes 1_{A^r}) = 1_{A^r}$$

We can think of ϵ^r_A and ϵ^l_A as cancellation and η^r_A and η^l_A as creation. The snake equations then say that given an object A in the category, if we create another object A and its dual A^l to the left of the original, and then cancel the original object with the dual we introduced, then this is the same as doing nothing at all to A – it is the identity map 1_A on A.

The dual objects are also represented as wires. To distinguish objects and their duals, wires are *directed*. The wire representing an object A is directed down the page, and wires representing dual objects A^r and A^l are directed up the page.

The ϵ and η maps are called **caps** and **cups** respectively, and are depicted graphically as:

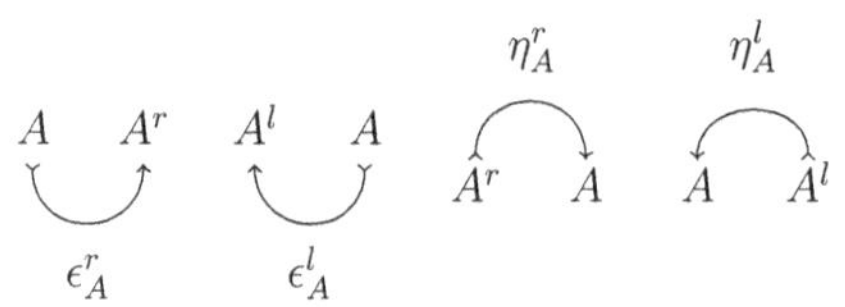

Graphically, the snake equations become

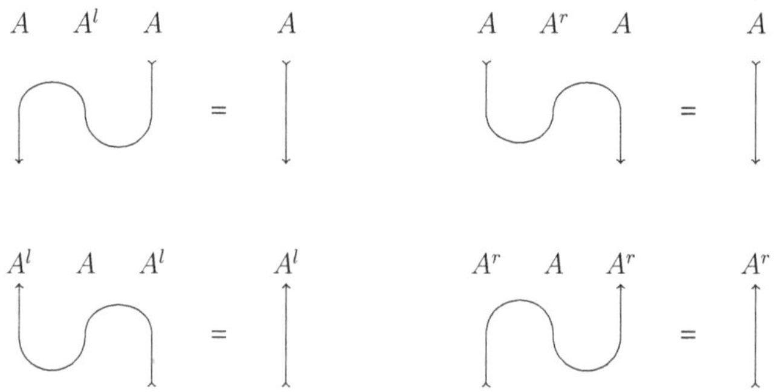

Compact closed categories are a convenient level of abstraction at which to work. Many of the categories one would think to use as either grammar or meaning categories have a compact closed structure.

Example 3 All objects in **Rel** are self-dual. Both caps are given by

$$\epsilon_X : X \otimes X \to \{\star\} :: \{((x, x), \star) \mid x \in X\}$$

The associated cup η_X is the converse of the above. The snake equations can be verified by direct calculation.

Example 4 **FHilb** is the category of finite dimensional real inner product spaces. As in the case of $\mathbf{FdVect}_{\mathbb{R}}$, the tensor $\otimes$ is the tensor product of vector spaces and I is the one-dimensional space $\mathbb{R}$. In defining cups and caps, we make use of the fact that if $\{v_i\}_i$ and $\{u_j\}_j$ are bases for vector spaces V and U respectively, then $\{v_i \otimes u_j\}_{i,j}$ is a basis for $V \otimes U$. Moreover, any linear map is fully determined by its action on a basis. Every finite-dimensional vector space is self-dual, and the cups and caps are given by

$$\epsilon_V : V \otimes V \to \mathbb{R} :: \sum_{i,j} c_{i,j}\, (v_i \otimes v_j) \mapsto \sum_{i,j} c_{i,j} \langle v_i | v_j \rangle$$

$$\eta_V : \mathbb{R} \to V \otimes V :: 1 \mapsto \sum_i (v_i \otimes v_i)$$

Verifying that these maps satisfy the snake equations again follows from a straight-forward calculation.

Remark 1 The tensor in a compact closed category is *not* necessarily a categorical product. Specifically, we cannot expect to describe every state of $A \otimes B$ as a tensor of two states taken from A and B. We can understand this as showing composite systems have interesting behaviour that cannot be explained in terms of the behaviour of their component parts. In fact, if the tensor of a compact closed category happens to be a categorical product, then that category must be trivial, in a precise mathematical sense (Coecke and Kissinger 2017).

9.2.3 Grammar Categories

Many algebraic gadgets exist to model grammar, as detailed in Coecke (2013) for example. In the present work, we use Lambek's pregroup grammars, as many grammars have a pregroup structure (Sadrzadeh 2007). Moreover, pregroups can be viewed as compact closed categories. It should be emphasized though that our approach does not depend on pregroups, and can be applied to other grammatical models. The definition of pregroups relies on the notion of a monoid, which is a set with associative multiplication and identity but no inverse.

Definition 3 A **pregroup** is a tuple $(P, \cdot, 1, -^l, -^r, \leq)$ where $(P, \cdot, 1, \leq)$ is a partially ordered monoid and $-^r, -^l$ are functions $P \to P$ such that $\forall p \in P$,

$$p \cdot p^r \leq 1 \tag{9.2}$$

$$p^l \cdot p \leq 1 \tag{9.3}$$

$$1 \leq p^r \cdot p \tag{9.4}$$

$$1 \leq p \cdot p^l \tag{9.5}$$

The $\cdot$ will usually be omitted, writing pq for $p \cdot q$. One interprets grammar by freely generating a pregroup from a set of atomic linguistic types. Words are then assigned an element of the pregroup depending on their linguistic function. A string of words is then mapped to an element of the pregroup by multiplying together the elements associated with its constituent words in their syntactic order. If $s_1 \cdots s_n \leq t$, we say that the type $s_1 \cdots s_n$ **reduces** to the type t. The pregroup freely generated by a set P is denoted by Preg_P.

Example 5 For simplicity, we only use the linguistic types n, for noun, and s, for sentence. Hence, we will work with $\mathsf{Preg}_{\{n,s\}}$. Consider the sentence *'Chickens cross roads.'* The nouns *chickens* and *roads* are of type n, and the transitive verb *cross* is assigned the type $n^r s n^l$. *'Chickens cross roads'* therefore has type $n(n^r s n^l)n$. Then we have the following type reductions:

$$\begin{aligned} n(n^r s n^l)n &= (nn^r)s(n^l n) \\ &\overset{(9.2)}{\leq} s(n^l n) \\ &\overset{(9.3)}{\leq} s \end{aligned}$$

Note that we could also have performed these two steps in the opposite order. The above reduction can be given a neat graphical interpretation as follows:

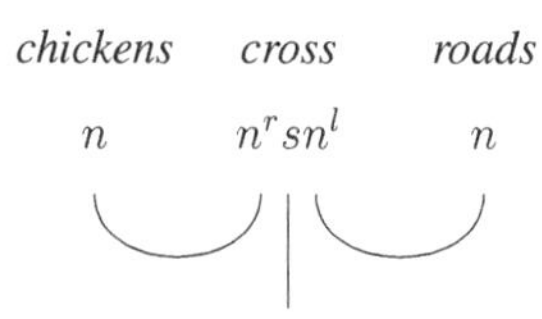

where it is now very clear that the order of the reductions doesn't matter. This is a feature that is typical for pregroup grammars, while other categorial grammars such as Lambek's original categorial grammar (Lambek 1958) have more constraints on the order of the reductions.

A pregroup can be considered as a compact closed category. The objects of this category are the elements of the pregroup. The morphisms are given by the order structure of the pregroup. That is, there is a unique morphism $p \to q$ if and only if $p \leq q$. The tensor $\otimes$ is the monoid multiplication and the monoidal unit is the element 1. Unsurprisingly, the left and right duals of p are p^l and p^r respectively. The cups and caps are the unique morphisms given by the pregroup axioms (9.2), (9.3), (9.4), and (9.5). In the reduction diagram, note how the cups correspond to the cups of the compact closed structure.

9.2.4 Meaning Categories

Distributional models of meaning use vector spaces to represent the meaning of words. In this paper we move away from this approach, choosing instead to work in **ConvexRel**, the category of convex sets and convexity-respecting relations. Before describing this new setting, we first present the conventional vector space approach as a demonstration of the categorical compositional method. This will prepare the ground for detailed discussion of **ConvexRel** to later sections.

One way to model meanings in a vector space is to use co-occurrence statistics (Bullinaria and Levy 2007). The meaning of a word is identified with the frequency with which it appears near other words. One first chooses a collection of *context words*. These will be the basis vectors. One then analyses a large corpus of writing to determine how often words co-occur with the context words. For example, suppose the word *dog* appears in the same context as *cat*, *companion*, and *cuisine* respectively 19, 25, and 2 times. If our collection of context words is {*cat*, *companion*, *cuisine*}, then *dog* would be assigned the vector $(19, 25, 2)$. A drawback of the co-occurrence approach is that antonyms appear in similar contexts and, hence, words such as ‘win’ and ‘lose’ are indistinguishable (despite evidently being different in meaning). Another related difficulty is that vector spaces are notoriously bad for representing basic propositional logic. Nonetheless, the vector space model is highly successful in NLP.

One can also use basis vectors to represent *quality dimensions*. For example, Rosch and Mervis (1975), Tversky (1977), and Hampton (1987) all represent concepts as feature vectors, with basis dimensions representing attributes of the concept. So, for instance, one might represent the word *dog* with certain values on the *fluffy*, *loyal*, and *wolf-like*. This approach closely resembles ours in this paper, though we move away from the vector space setting.

9.2.5 Putting it All Together

We now have all the necessary components: a grammar category and a meaning category. We used the examples of pregroup grammars and vector spaces, but we stress yet again that the abstract method applies equally well to any two compact closed categories.

Suppose we have a string of words $w_1 \cdots w_n$ and a pregroup P. Suppose further that $w_i \in W_i$, where W_i is the vector space associated with w_i's linguistic type. The meaning of $w_1 \cdots w_n$ is computed as follows:

1. Assign a pregroup element p_i to each word w_i based on its linguistic type.
2. Apply pregroup reduction rules (cups and caps) to the element $p_1 p_2 \cdots p_n$ to obtain a simpler type x such that:

$$p_1 p_2 \cdots p_n \leq x.$$

3. Thinking of the above reduction as a morphism in the pregroup built up from ϵ, η and identities (as given by its reduction diagram), apply the corresponding vector space morphism of type:

$$W_1 \otimes \cdots \otimes W_n \to W_X$$

to the string of word meanings represented as $w_1 \otimes \cdots \otimes w_n$. the vector space morphism will be a combination of ϵ, η and identity morphisms, where the ϵ, η morphisms are as described in Example 4.

Example 6 Consider again the sentence *'chickens cross roads'*. The nouns *chickens* and *roads* have type n and so are represented in some vector space N of nouns. The transitive verb *cross* has type $n^r s n^l$ and, hence, is represented by a vector in the vector space $N \otimes S \otimes N$ where S is a vector space modelling sentence meaning. The meaning of *'chickens cross roads'* is the image of

$$\overrightarrow{chickens} \otimes \overrightarrow{cross} \otimes \overrightarrow{roads} \tag{9.6}$$

under the map

$$\epsilon_N \otimes 1_S \otimes \epsilon_N : N \otimes (N \otimes S \otimes N) \otimes N \to S \tag{9.7}$$

This nicely illustrates the general method. Our meaning category supplies the qualitative meanings of *chickens*, *cross*, and *roads*. Our grammar category then tells us how to stitch these together. This corresponds to 'telling us where to put cups and caps.' The essence of the method should be thought of as the diagram

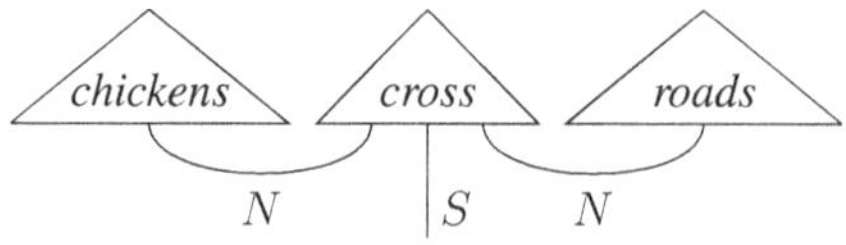

where we think of the words as meaning vectors (9.6) and the wires as the map (9.7). In fact, rather than just using cups and identity wires, which are enough to account for the grammatical structure captured by pregroups, we can enrich the graphical language to directly account for meanings of functional words, such as relative pronouns.

9.2.6 Beyond Standard Categorial Grammar

A first rather trivial example of a functional word is "does", which can be accounted for by means of caps (Coecke et al. 2010):

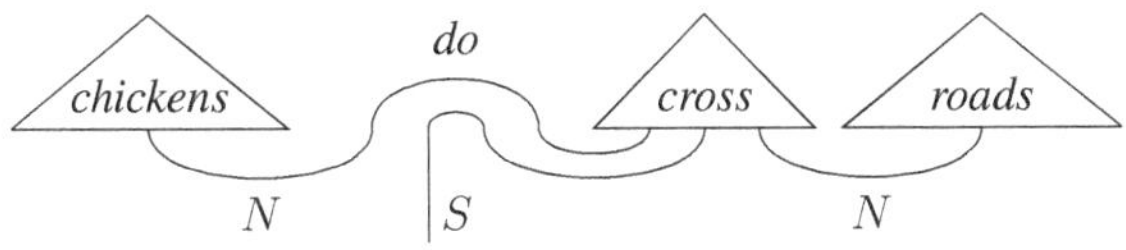

Things become more interesting when we introduce, rather than just wires, the idea of a *multi-wire* (also called *spider* (Coecke and Kissinger 2017)) which can have more than two ends, or fewer:

$\mu_N :=$ 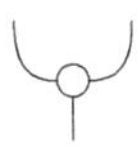 $\iota_S :=$ 

The way these behave is just like wires. The only thing that matters is: 'either being connected by a multi-wire, or not'. As a consequence, multi-wires 'fuse' together:

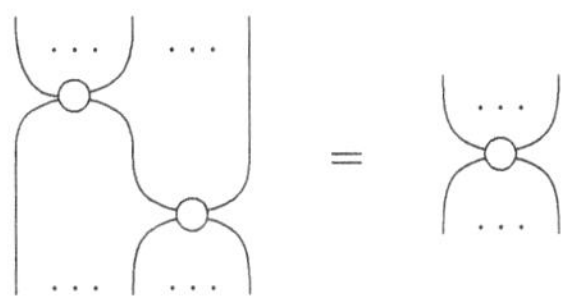

These multi-wires can be defined in category-theoretic terms for any symmetric monoidal category, where they are called *commutative special dagger Frobenius structures* (Coecke and Kissinger 2017; Coecke et al. 2013b).

Example 7 On each set X in the category **Rel** one can take the relations

$$X \otimes \ldots \otimes X \to X \otimes \ldots \otimes X :: \{((x, \ldots, x), (x, \ldots, x)) \mid x \in X\}$$

to be the multi-wires, and note in particular that these include identities, cups and caps.

Example 8 On each inner product space V in the category $\mathbf{FdVect}_{\mathbb{R}}$, given any orthonormal basis $\{v_i\}_i$ for V, one can take the linear maps

$$V \otimes \ldots \otimes V \to V \otimes \ldots \otimes V :: v_{i_1} \otimes \ldots \otimes v_{i_n} \mapsto \delta_{i_1 \ldots i_n} v_{i_1} \otimes \ldots \otimes v_{i_1}$$

to be the multi-wires, and again these include identities, cups and caps.

Using these multi-wires we can now express the meaning of relative pronouns (Sadrzadeh et al. 2013, 2016):

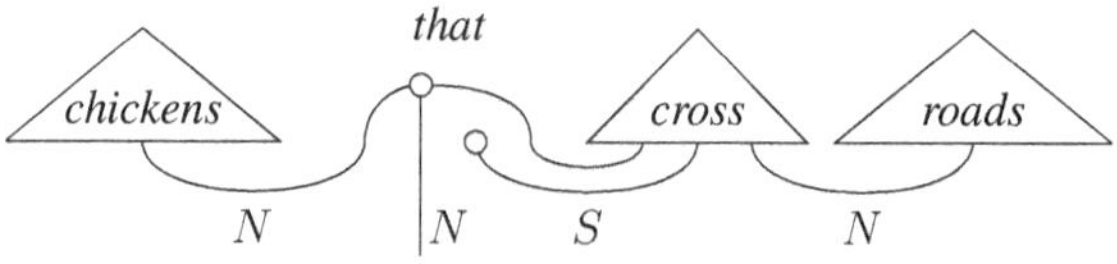

Firstly, note that what we obtain is a noun rather than a sentence, and one that is closely related to 'dead chicken'. Secondly, note that the use of the three-wire is mainly conjunction, conjoining '[is] chicken' and 'crosses road'. In this paper we will use them to directly express conjunctions. The one-wire gets rid of the sentence type.

9.3 Conceptual Spaces

Conceptual spaces are proposed in Gärdenfors (2004) as a framework for representing information at the conceptual level. Gärdenfors contrasts his theory with both a symbolic approach to concepts, and an associationist approach where concepts are represented as associations between different kinds of information elements. Instead, conceptual spaces are structures based on quality dimensions such as weight, height, hue and brightness. Conceptual spaces have an internal structure based on how quality dimensions interact with each other. A pair (or set) of dimensions is called *integral* if assignment of a value on one dimension requires assignment of a value on another dimension. For example, the dimensions of hue, saturation, and value in the HSV colour space are integral. Dimensions are called *separable* if values on one dimension can be assigned independently from the others. For example, hue and height are separable. A set of integral dimensions is called a domain, and a conceptual space is composed of a number of domains linked in some way. A concept then corresponds to a convex region in a conceptual space. The way in which the interaction between quality dimensions is specified in Gärdenfors's model is by using different distance metrics. Within a set of integral dimensions, distance is Euclidean, and between sets of integral dimensions, the city block metric is used.

Concept composition within conceptual spaces has been formalized in Rickard et al. (2007), Adams and Raubal (2009), and Lewis and Lawry (2016) for example. Many of these approaches focus on noun-noun composition, rather than utilising any more complex structure, and the way in which nouns compose often focuses on correlations between attributes in concepts. Since then, Gärdenfors has started to formalise verb spaces, adjectives, and other linguistic structures (Gärdenfors 2014). However, a systematic method for how to utilise grammatical structures within conceptual spaces has not yet been provided. In this sense, the category-theoretic approach to concept composition we describe below will introduce a more general approach to concept composition that can apply to varying grammatical types.

9.4 The Category of Convex Relations

In NLP applications, meanings are typically interpreted in categories of real vector spaces. For our intended cognitive application, we now introduce a category that emphasizes convex structure. The familiar definition of convex set is a subset of a vector space which is closed under forming convex combinations. In this paper we consider a more general setting that includes convex subsets of vector spaces, but also allows us to consider some further, more discrete, examples.

We begin with some convenient notation. For a set X we write $\sum_i p_i |x_i\rangle$ for a finite formal convex sum of elements of X, where $p_i \in \mathbb{R}^{\geq 0}$ and $\sum_i p_i = 1$. We then write $D(X)$ for the set of all such sums. Here we abuse the physicists' ket notation to highlight that our sums are formal, following a convention introduced in Jacobs (2011). Equivalently, these sums can be thought of as finite probability distributions on the elements of X.

A **convex algebra** is a set A with a function $\alpha : D(A) \to A$ satisfying the following conditions:

$$\alpha(|a\rangle) = a \qquad \text{and} \qquad \alpha\left(\sum_{i,j} p_i q_{i,j} |a_{i,j}\rangle\right) = \alpha\left(\sum_i p_i |\alpha(\sum_j q_{i,j} |a_{i,j}\rangle)\rangle\right) \tag{9.8}$$

Informally, α is a **mixing operation** that allows us to form convex combinations of elements, and the equations in (9.8) require the following good behaviour:

- Forming a convex combination of a single element a returns a as we would expect
- Iterating forming convex combinations interacts with flattening sums of sums in the way we would expect

We consider some examples of convex algebras.

Example 9 The closed real interval [0, 1] has an obvious convex algebra structure. Similarly, every real or complex vector space has a natural convex algebra structure using the underlying linear structure.

Example 10 (Simplices) For any set X, the formal convex sums of elements of X themselves form the **free convex algebra** on X, which can also be seen as a simplex with vertices the elements of X. Mixtures are formed as follows:

$$\sum_i p_i | \sum_j q_{i,j} |x_{i,j}\rangle\rangle \mapsto \sum_{i,j} p_i q_{i,j} |x_{i,j}\rangle$$

Example 11 (Semilattices) As a slightly less straightforward example, every affine join semilattice (that is, one that has all finite *non-empty* joins) has a convex algebra structure given by:

$$\sum_i p_i |a_i\rangle = \bigvee_i \{a_i \mid p_i > 0\}$$

Notice that here the scalars p_i are discarded and play no active role. These "discrete" types of convex algebras allow us to consider objects such as the Boolean truth values.

Example 12 (Trees) Given a finite tree, perhaps describing some hierarchical structure, we can construct an affine semilattice in a natural way. For example, consider a limited universe of foods, consisting of bananas, apples, and beer. Given two members of the hierarchy, their join will be the lowest level of the hierarchy which is above them both. For instance, the join of *bananas* and *apples* would be *fruit*.

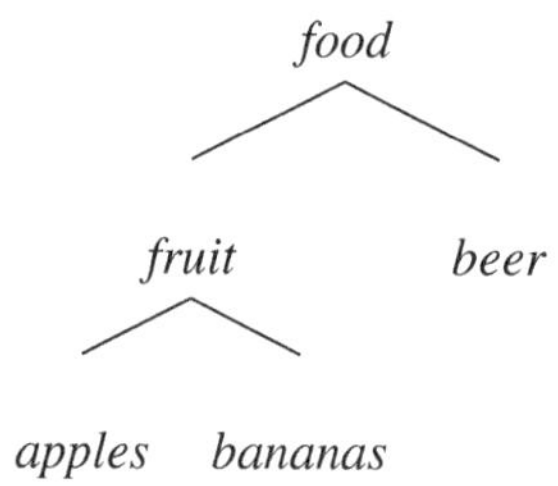

When α can be understood from the context, we abbreviate our notation for convex combinations by writing:

$$\sum_i p_i a_i := \alpha(\sum_i p_i |a_i\rangle)$$

Using this convention, we define a **convex relation** of type $(A, \alpha) \to (B, \beta)$ as a binary relation $R : A \to B$ between the underlying sets that commutes with forming convex mixtures as follows:

$$(\forall i. R(a_i, b_i)) \Rightarrow R\left(\sum_i p_i a_i, \sum_i p_i b_i\right)$$

We note that identity relations are convex, and convex relations are closed under relational composition and converse.

Example 13 (Homomorphisms) If (A, α) and (B, β) are convex algebras, functions $f : A \to B$ satisfying:

$$f(\sum_i p_i x_i) = \sum_i p_i f(x_i)$$

are convex relations. These functions are the **homomorphisms of convex algebras**. The identity function and constant functions are examples of homomorphisms of convex algebras.

The singleton set $\{*\}$ has a unique convex algebra structure, denoted I. Convex relations of the form $I \to (A, \alpha)$ correspond to **convex subsets**, that is, subsets of A closed under forming convex combinations.

Definition 4 We define the category **ConvexRel** as having convex algebras as objects and convex relations as morphisms, with composition and identities as for ordinary binary relations.

Given a pair of convex algebras (A, α) and (B, β) we can form a new convex algebra on the cartesian product $A \times B$, denoted $(A, \alpha) \otimes (B, \beta)$, with mixing operation:

$$\sum_i p_i |(a_i, b_i)\rangle \mapsto \left(\sum_i p_i a_i, \sum_i p_i b_i\right)$$

This induces a symmetric monoidal structure on **ConvexRel**. In fact, the category **ConvexRel** has the necessary categorical structure for categorical compositional semantics:

Theorem 1 *The category* **ConvexRel** *is a compact closed category. The symmetric monoidal structure is given by the unit and monoidal product outlined above. The caps for an object* (A, α) *are given by:*

$$\frown : I \to (A, \alpha) \otimes (A, \alpha) :: \{(*, (a, a)) \mid a \in A\}$$

the cups by:

$$\smile : (A, \alpha) \otimes (A, \alpha) \to I :: \{((a, a), *) \mid a \in A\}$$

and more generally, the multi-wires by:

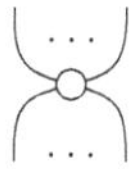
$$: A \otimes \ldots \otimes A \to A \otimes \ldots \otimes A :: \{((a, \ldots, a), (a, \ldots, a)) \mid a \in A\}$$

Note that in the definition of the multi-wires and for the remainder of the paper, we abuse notation and leave the algebra α on A implicit.

Remark 2 In particular, the multi-wires μ_A and ι_A are defined as follows:

$$\mu_A : A \otimes A \to A :: \{((a, a), a) | a \in A\} \qquad \iota_A : A \to I :: \{(a, *) | a \in A\}$$

Remark 3 As observed in Remark 1, as **ConvexRel** is compact closed, its tensor cannot be a categorical product. For example, there are convex subsets of $[0, 1] \otimes [0, 1]$ such as the diagonal:

$$\{(x, x) \mid x \in [0, 1]\}$$

that cannot be written as the cartesian product of two convex subsets of $[0, 1]$. This behaviour exhibits non-trivial *correlations* between the different components of the composite convex algebra.

Remark 4 We have given an elementary description of **ConvexRel**. More abstractly, it can be seen as the category of relations in the Eilenberg-Moore category of the finite distribution monad. Its compact closed structure then follows from general principles (Carboni and Walters 1987).

9.5 Noun, Adjective, and Verb Concepts

We define a **conceptual space** to be an object of **ConvexRel**. In order to match the structure of the pregroup grammar, we require two distinct objects: a noun space N and a sentence space S.

The **noun space** N is given by a composite

$$N_{colour} \otimes N_{taste} \otimes \ldots$$

describing different attributes such as colour and taste. A **noun** is then a convex subset of such a space. In our examples, we take our sentence space to be a convex algebra in which the individual points are events. Our general scheme can incorporate other sentence space structures, such choices are generally specific to the application under consideration. A **sentence** is then a convex subset of S.

We now describe some example noun and sentence spaces. We then show how these can be combined to form spaces describing adjectives and verbs. Once we have these types available, we show in Sect. 9.6 how concepts interact within sentences.

9.5.1 Example: Food and Drink

We consider a conceptual space for food and drink as our running example. The space N is composed of the domains N_{colour}, N_{taste}, $N_{texture}$, so that

$$N = N_{colour} \otimes N_{taste} \otimes N_{texture}$$

The domain N_{colour} is the RGB colour domain, i.e. triples $(R, G, B) \in [0, 1]^3$ with R, G, B standing for intensity of red, green, and blue light respectively. N_{taste} is defined as the simplex of combinations of four tastes: sweet, sour, bitter, and salt. We therefore have

$$N_{taste} = \{\vec{t} | \vec{t} = \sum_{i \in I} w_i \vec{t}_i\} \tag{9.9}$$

where $I = \{sweet, sour, bitter, salt\}$, $\vec{t}_i$ is the vector in some chosen basis of $\mathbb{R}^4$ whose elements are all zero except for the ith element whose value is one, and $\sum_i w_i = 1$. $N_{texture}$ is just the set $[0, 1]$ ranging from completely liquid (0) to completely solid (1). We define a property $p_{property}$ to be a convex subset of a domain, and specify the following examples (see Figs. 9.1 and 9.2):

$$p_{yellow} = \{(R, G, B) | (R \geq 0.7), (G \geq 0.7), (B \leq 0.5)\}$$

$$p_{green} = \{(R, G, B) | (R \leq G), (B \leq G), (R \leq 0.7), (B \leq 0.7), (G \geq 0.3)\}$$

$$p_{sweet} = \{\vec{t} | t_{sweet} \geq t_l \text{ for } l \neq sweet\}$$

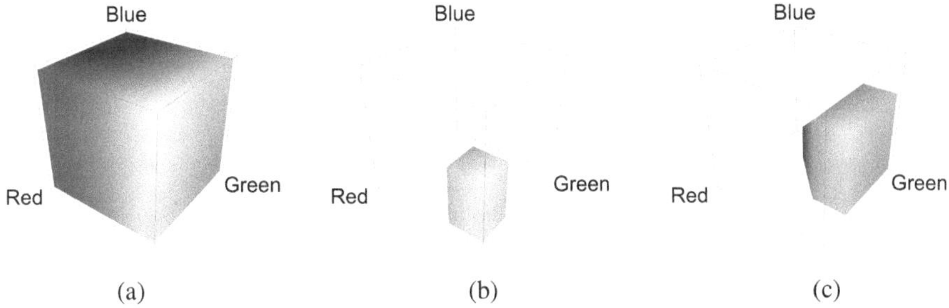

Fig. 9.1 The RGB colour cube and properties p_{colour}. (**a**) The RGB colour cube. (**b**) Property p_{yellow}. (**c**) Property p_{green}

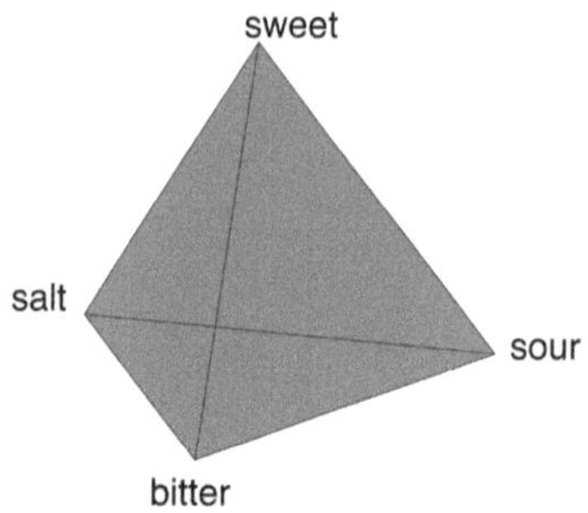

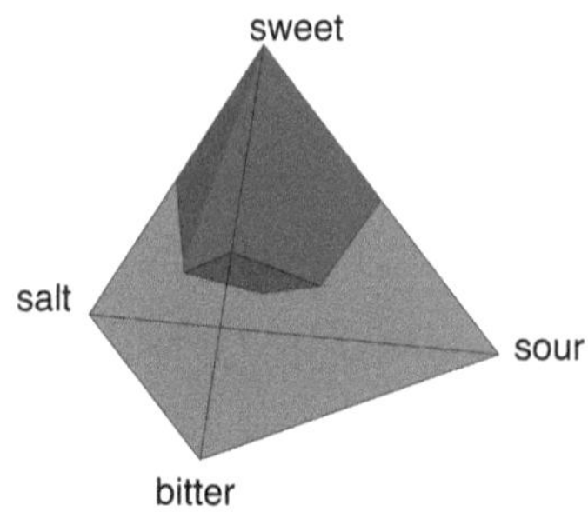

Fig. 9.2 The taste space and the property p_{sweet}. (**a**) The taste tetrahedron. (**b**) The region corresponding to $p_{sweet} = \{\vec{t} | t_{sweet} \geq t_l \text{ for } l \neq sweet\}$

The properties p_{sour} and p_{bitter} are defined analogously.

9.5.1.1 Nouns

We define some nouns below. Properties in the colour domain are specified using sets of linear inequalities, and colours in the taste domain are specified using the convex hull of sets of points. We use $Conv(A)$ to refer to the convex hull of a set A.

$$\begin{aligned}
banana &= \{(R, G, B) | (0.9R \leq G \leq 1.5R), (R \geq 0.3), (B \leq 0.1)\} \\
&\quad \otimes Conv(\{t_{sweet}, 0.25t_{sweet}{+}0.75t_{bitter}, 0.7t_{sweet}{+}0.3t_{sour}\}) \otimes [0.2, 0.5] \\
apple &= \{(R, G, B) | (R - 0.7 \leq G \leq R + 0.7), (G \geq 1 - R), (B \leq 0.1)\} \\
&\quad \otimes [0.5, 1] \otimes Conv(\{t_{sweet}, 0.75t_{sweet} + 0.25t_{bitter}, 0.3t_{sweet} + 0.7t_{sour}\}) \\
&\quad \otimes [0.5, 0.8] \\
beer &= \{(R, G, B) | (0.5R \leq G \leq R), (G \leq 1.5 - 0.8R), (B \leq 0.1)\} \\
&\quad \otimes Conv(\{t_{bitter}, 0.7t_{sweet} + 0.3t_{bitter}, 0.6t_{sour} + 0.4t_{bitter}\}) \otimes [0, 0.01]
\end{aligned}$$

where t_i are as given in (9.9). The tensor product $\otimes$ used in these equations is the tensor product in **ConvexRel**, and is therefore the Cartesian product of sets.

The subsets of points representing tastes are explained as follows using the case of banana as an example. Bananas are not at all salty, and therefore w_{salt} is set to 0. Bananas are sweet, and therefore the point t_{sweet} is chosen as an extremal point in the set of banana tastes. Bananas can also be somewhat but not totally bitter, and therefore the point $0.25t_{sweet} + 0.75t_{bitter}$ is chosen as an extremal point. Similarly bananas can be a little sour, and therefore $0.7t_{sweet} + 0.3t_{sour}$ is also chosen as an extremal point. Finally the convex hull of these points is formed giving a set of points corresponding to banana taste.

Pictorially, we have:

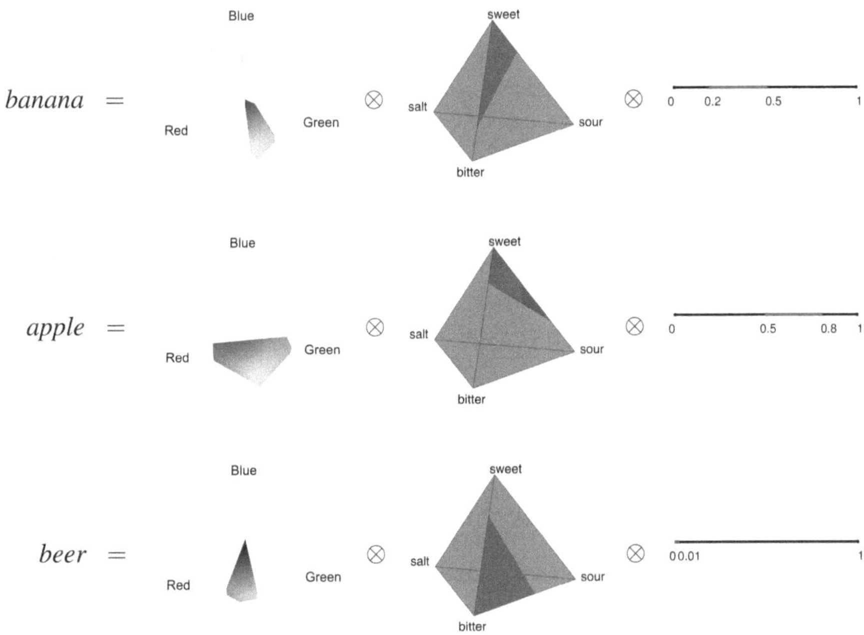

What is an appropriate choice of sentence space for describing food and drink? We need to describe the events associated with eating and drinking. We choose a very simple structure where the events are either positive or negative, and surprising or unsurprising. We therefore use a sentence space of pairs. The first element of the pair states whether the sentence is positive (1) or negative (0) and the second states whether it is surprising (1) or unsurprising (0). The convex structure on this space is the convex algebra on a join semilattice induced by element-wise max, as in Example 11. We therefore have four points in the space: positive, surprising (1, 1); positive, unsurprising (1, 0); negative, surprising (0, 1); and negative, unsurprising (0, 0). Sentence meanings are convex subsets of this space, so they could be singletons, or larger subsets such as $negative = \{(0,\ 1), (0,\ 0)\}$.

9.5.1.2 Adjectives

Recall that in a pregroup setting the adjective type is nn^l. In **ConvexRel**, the adjective therefore has type $N \otimes N$. Adjectives are convex relations on the noun space, so can be written as sets of ordered pairs. We give two examples, $yellow_{adj}$ and $soft_{adj}$. The adjective $yellow_{adj}$ has the simple form:

$$\{(\overrightarrow{x}, \overrightarrow{x})|x_{colour} \in p_{yellow}\}$$

This simple form reflects the fact that $yellow_{adj}$ depends only on one area of the conceptual space, so it really just corresponds to the property p_{yellow}.

An adjective such as 'soft' behaves differently to this. We cannot simply define soft as one area of the conceptual space, because whether or not something is soft depends what it was originally. Using relations, we can start to write down the right type of structure for the adjective, as long as the objects are sufficiently distinct. Restricting our universe just to bananas and apples, we can write $soft_{adj}$ as

$$\{(\overrightarrow{x}, \overrightarrow{x})|\overrightarrow{x} \in banana \text{ and } x_{texture} \leq 0.35 \text{ or } \overrightarrow{x} \in apples \text{ and } x_{texture} \leq 0.6\}$$

Note that here, we are using *banana* and *apple* as shorthand for specifications of convex areas of the conceptual space. These could be written out in longhand as sets of inequalities within the colour and taste spaces.

An analysis of the difficulties in dealing with adjectives set-theoretically, breaking them down into (roughly) three categories, is given in Kamp and Partee (1995). Under this view, both adjectives and nouns are viewed as one-place predicates, so that, for example $red = \{x|x \text{ is red}\}$ and $dog = \{x|x \text{ is a dog}\}$. There are then three classes of adjective. For **intersective** adjectives, the meaning of *adj noun* is given by $adj \cap noun$. For **subsective** adjectives, the meaning of *adj noun* is a subset of *noun*. For **privative** adjectives, however, $adj\ noun \not\subseteq noun$.

Intersective adjectives are a simple modifier that can be thought of as the intersection between two concepts. We can make explicit the internal structure of these adjectives exploiting the multi-wires of Theorem 1. For example, in the case of *yellow banana*, we take the intersection of *yellow* and *banana*. We then show how to understand *yellow* as an adjective. While the general case of adjectives is depicted as:

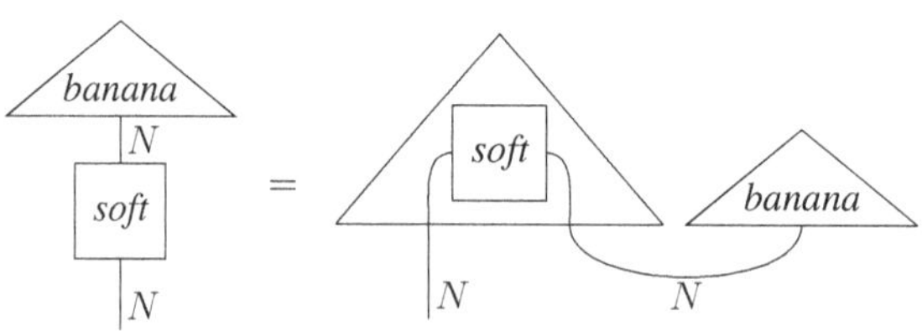

in the case of intersective adjectives the diagrams specialise to:

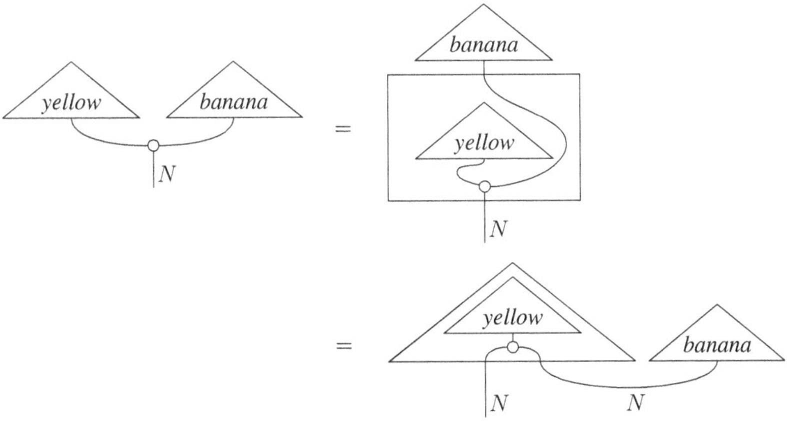

This shows us how the internal structure of an intersective adjective is derived directly from a noun.

9.5.1.3 Verbs

The pregroup type for a transitive verb is $n^r s n^l$, mapping to $N \otimes S \otimes N$ in **ConvexRel**. To define the verb, we use concept names as shorthand, where these can easily be calculated. For example, is *green* is considered to be an intersective adjective, *green banana* can be calculated by taking the intersection of *green* and *banana* by combining the inequalities specifying the colour property, giving:

$$
\begin{aligned}
\textit{green banana} = \{(R, G, B) | (R \leq G \leq 1.5R), (G \leq B), (0.3 \leq R \leq 0.7), (G \leq 0.3)\} \\
\otimes\, Conv(\{t_{sweet}, 0.25t_{sweet} + 0.75t_{bitter}, 0.7t_{sweet} + 0.3t_{sour}\}) \\
\otimes\, [0.2, 0.5]
\end{aligned}
$$

Although a full specification of a verb would take in all the nouns it could possibly apply to, for expository purposes we restrict our nouns to just bananas and beer which do not overlap, due to the fact that they have different textures. We define the verb *taste* : $I \to N \otimes S \otimes N$ as follows:

$$
\begin{aligned}
\textit{taste} = (\textit{green banana} \otimes \{(0, 0)\} \otimes \textit{bitter}) \cup (\textit{green banana} \otimes \{(1, 1)\} \otimes \textit{sweet}) \\
\cup\, (\textit{yellow banana} \otimes \{(1, 0)\} \otimes \textit{sweet}) \cup (\textit{beer} \otimes \{(0, 1)\} \otimes \textit{sweet}) \\
\cup\, (\textit{beer} \otimes \{(1, 0)\} \otimes \textit{bitter})
\end{aligned}
$$

The full specification of a verb would take in all the nouns it could apply to. However, this does not mean that each noun must be specified separately. Some algebraic description of the relation could be used to describe how each point in the space is affected.

9.5.2 Example: Robot Movement

We now present another example describing a simple formulation of robot movement. We will describe our choices of noun space N and sentence space S, and show how to form nouns and verbs.

9.5.2.1 Nouns

The types of nouns we wish to describe are objects, such as *armchair* and *ball*, the robots *Cathy* and *David*, and places such as *kitchen* and *living room*. For shorthand, we call these nouns a, b, c, d, k, and l. These are specified in the noun space N which is itself composed of a number of domains

$$N_{location} \otimes N_{direction} \otimes N_{shape} \otimes N_{size} \otimes N_{colour} \otimes \ldots$$

We firstly consider the kitchen and living rooms as being defined by convex subsets of points in the domain $N_{location}$, defining properties in the location domain as:

$$p_{kitchen\,location} = \{(x_1, x_2)|x_1 \in [0, 5], x_2 \in [0, 10]\}$$
$$p_{living\,room\,location} = \{(x_1, x_2)|x_1 \in [5, 10], x_2 \in [0, 10]\}$$

which can be depicted as follows:

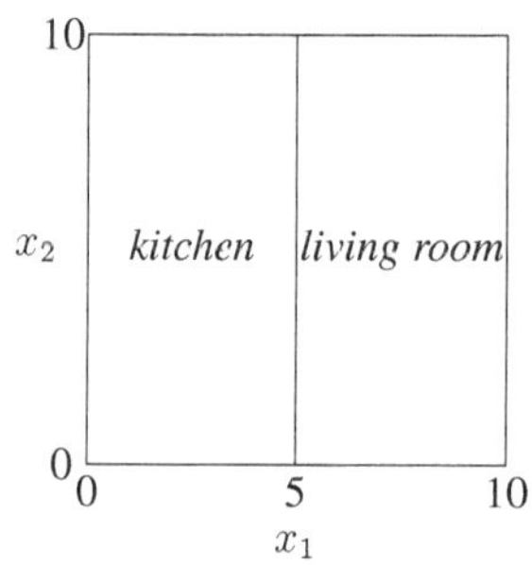

Then the nouns *kitchen* and *living room* are given by these properties together with other sets of characteristics in the shape domain, size domain, and so on, which we won't specify here.

$$kitchen = p_{kitchen\,location} \otimes p_{kitchen\,shape} \otimes p_{kitchen\,size} \otimes \ldots$$
$$living\ room = p_{living\,room\,location} \otimes p_{living\,room\,shape} \otimes p_{living\,room\,size} \otimes \ldots$$

Similarly, the other nouns are defined by combinations of properties in the noun space. For this example, we do not worry too much about what they are, but assume that they allow us to distinguish between the objects.

9.5.2.2 Verbs

In order to define some verbs, we need consider what a suitable sentence space should look like. We want to give sentences of the form:

The ball is in the living room
Cathy moves to the kitchen

In these sentences, an object or agent is related to a path through time and space. Note that in the case of the verb 'is in', this path is in fact trivially just a point, however for 'moves to', the path actually is a path through time and space, and we will need to use subsets of the time and location domains to specify one single event. We therefore define the sentence space to be comprised of the noun space N, a time dimension T, and the location domain $N_{location}$:

$$S = N \otimes T \otimes N_{location}$$

The agent is represented by a point in the noun space N, and the path they take as described in the sentence is represented as a subset of the time and location domains. In what follows, we think of 0 on the time dimension T as referring to 'now', with negative values of T referring to the past and positive values referring to the future.

As in the food example, transitive verbs are of the form $N \otimes S \otimes N$. This means that in this example, they are of the form

$$N \otimes N \otimes T \otimes N_{location} \otimes N,$$

and can be thought of as sets of ordered tuples of the form

$$(n_1, n_2, t, l, n_3),$$

where n_i stands for points in the noun space, t is a time, l is a location. We will consider the following verbs: *isin*, *movesto*.[1] The verb *isin* can take any of the nouns

[1]It could be argued that these are not transitive verbs, but intransitive verbs plus preposition. However, we can parse the combination as a transitive verb, since a preposition has type $s^r s n^l$ and therefore the combination reduces to type of a transitive verb:

$$(n^r s)(s^r s n^l) \leq n^r s n^l$$

a, b, c, or d as subject, and any of k, l as object. This verb refers to just one timepoint, i.e. now, or 0. The verb is as follows:

$$isin = \{(\vec{n}, \vec{n}, t_{now}, m_{location}, \vec{m}) | \vec{n} \in a \cup b \cup c \cup d, t_{now} = 0, \vec{m} \in k \cup l\} \quad (9.10)$$

The verb *movesto* refers to more than one point in time. We need to talk about an object moving from being at one location at a past time, to another location at time 0, or now. This movement should be continuous, since the objects we are talking about do not teleport from one point to another. We will also restrict the subject of the sentence to being one of the nouns a, b, c, or d, as we don't want to talk about the kitchen and living rooms moving at this point. The object of the verb, however, can be any of the nouns, so we can say, for example, that 'Cathy moves to the armchair', or 'The ball moves to Dave' (presumably because Cathy kicked it). The most specific event that can be described in the space will track the exact path that an object takes through space and time. The meaning of a less specific sentence will be a convex subset of these trajectories. We now define the verb as follows:

$$\begin{aligned} movesto = \{&(\vec{n}, \vec{n}, [t, 0], f([t, 0]), \vec{m}) \mid \\ &\vec{n} \in a \cup b \cup c \cup d, t<0, f \text{ continuous }, f(t) \in \vec{n}_{location}, f(0) \in \vec{m}_{location}\} \end{aligned} \quad (9.11)$$

The constraints on t ensure that the movement happened in the past, and the constraints on f ensure that the movement is from the location of the subject to the location of the object of the verb. These definitions are a little complex, but we will see how they work in interaction. Note that the location nouns *kitchen* and *living room* might seem to be of a different type to the object and agent nouns *armchair*, *ball*, *Cathy* and *David*. For example, we have specified that *kitchen* and *livingroom* do not move around. In future research we will be extending to a richer type system which can take account of these sorts of differences, and which will in fact be closer to that proposed by Gärdenfors in (2014).

9.6 Concepts in Interaction

We have given descriptions of how to form the different word types within our model of categorical conceptual spaces. In this section we show how to apply the type reductions of the pregroup grammar within the conceptual spaces formalism.

9.6.1 Sentences in the Food Space

The application of $yellow_{adj}$ to *banana* works as follows.

$$
\begin{aligned}
\textit{yellow banana} &= (1_N \otimes \epsilon_N)(\textit{yellow}_{adj} \otimes \textit{banana}) \\
&= (1_N \otimes \epsilon_N)\{(\overrightarrow{x}, \overrightarrow{x})|x_{colour} \in \textit{yellow}\} \\
&\quad \otimes (\{(R, G, B)|(0.9R \le G \le 1.5R), (R \ge 0.3), (B \le 0.1)\} \\
&\quad \otimes Conv(\{t_{sweet}, 0.25t_{sweet} + 0.75t_{bitter}, 0.7t_{sweet} + 0.3t_{sour}\}) \\
&\quad \otimes [0.2, 0.5]) \\
&= \{(R, G, B)|(0.9R \le G \le 1.5R), (R \ge 0.7), (G \ge 0.7), (B \le 0.1)\} \\
&\quad \otimes Conv(\{t_{sweet}, 0.25t_{sweet} + 0.75t_{bitter}, 0.7t_{sweet} + 0.3t_{sour}\}) \\
&\quad \otimes [0.2, 0.5]
\end{aligned}
$$

Notice, in the last line, how the colour property has altered. This alteration restricts to yellow hues. This assumes that we can tell bananas and apples apart by shape, colour and so on. Then the same calculation gives us

$$
\begin{aligned}
\textit{soft apple} &= \{(R, G, B)|(R - 0.7 \le G \le R + 0.7), (G \ge 1 - R), (B \le 0.1)\} \\
&\quad \otimes Conv(\{t_{sweet}, 0.75t_{sweet}+0.25t_{bitter}, 0.3t_{sweet}+0.7t_{sour}\})\otimes[0.4, 0.6]
\end{aligned}
$$

Using the definition of *taste* that we gave, we find that although sweet bananas are good:

$$
\begin{aligned}
&\textit{bananas taste sweet} = (\epsilon_N \otimes 1_S \otimes \epsilon_N)(\textit{bananas} \otimes \textit{taste} \otimes \textit{sweet}) \\
&= (\epsilon_N \otimes 1_S)(\textit{banana} \otimes (\textit{green banana} \otimes \{(1, 1)\} \cup \textit{yellow banana} \otimes \{(1, 0)\})) \\
&= \{(1, 1), (1, 0)\} \\
&= \textit{positive}
\end{aligned}
$$

sweet beer is not so desirable:

$$
\begin{aligned}
\textit{beer tastes sweet} &= (\epsilon_N \otimes 1_S \otimes \epsilon_N)(\textit{beer} \otimes \textit{taste} \otimes \textit{sweet}) \\
&= \{(0, 1)\} \\
&= \textit{negative and surprising}
\end{aligned}
$$

The compositional semantics we use can also deal with relative pronouns, described in detail in Kartsaklis et al. (2013). Relative pronouns are words such as ‘which’. For example, we can form the noun phrase *Fruit which tastes bitter*. This has the following structure:

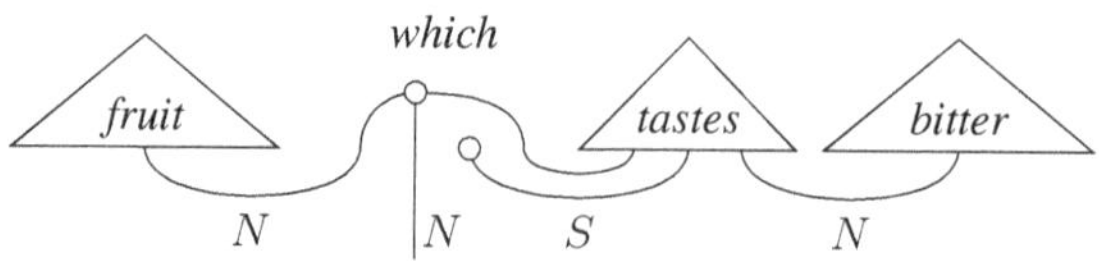

which simplifies to:

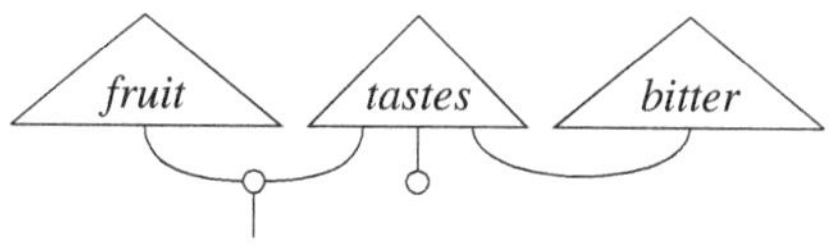

In our example, we find that *Fruit which tastes bitter* = *green banana*:

$$\begin{aligned}
\textit{Fruit which tastes bitter} &= (\mu_N \otimes \iota_S \otimes \epsilon_N)(Conv(\textit{bananas} \cup \textit{apples}) \otimes \textit{taste} \otimes \textit{bitter}) \\
&= (\mu_N \otimes \iota_S)(Conv(\textit{bananas} \cup \textit{apples}) \\
&\quad \otimes (\textit{green banana} \otimes \{(0, 0)\})) \\
&= \mu_N(Conv(\textit{bananas} \cup \textit{apples}) \otimes (\textit{green banana})) \\
&= \textit{green banana}
\end{aligned}$$

where μ_N is the Frobenius merge map on N and ι_S is the delete map on S described in Theorem 1 and Remark 2.

9.6.2 Sentences About Robot Movement

In this section we describe how to compute the meaning of sentences about robot movement. Our first example is the sentence 'Cathy moves to the living room'. In order to compute the meaning of this sentence, we assume that Cathy has a location.

$$\begin{aligned}
&\textit{Cathy moves to the living room} \\
&= (\epsilon_N \otimes 1_s \otimes \epsilon_N)(C \otimes \textit{movesto} \otimes L) \\
&= (\epsilon_N \otimes 1_s \otimes)(C \otimes \{(\vec{n}, \vec{n}, [t, 0], f([t, 0])) | f(0) \in L_{location}\}) \qquad (9.12) \\
&= \{C, [t, 0], f([t, 0]) | f(t) \in C_{location}, f(0) \in L_{location}\}
\end{aligned}$$

In line (9.12) further constraints apply to t and f as described in Eq. (9.11). This calculation gives us a set of continuous line segments starting from Cathy's location at time t and ending in the living room at time 0.

We now need to check that this set of line segments is convex. We assume that Cathy can take any possible location, and her other attributes remain static.

This means that the set of possible instantiations of Cathy is convex. The set of time segments $[t, 0]$ such that $t < 0$ forms a convex set. Consider two such time segments. We define a convex mixture of these segments pointwise:

$$p[t_1, 0] + (1 - p)[t_2, 0] = [pt_1 + (1 - p)t_2, p0 + (1 - p)0] = [pt_1 + (1 - p)t_2, 0]$$

which clearly satisfies the condition that the start point is in the past and the end point is now.

We then consider the convex mixture of two sets of locations $f_1([t_1, 0])$ and $f_2([t_2, 0])$. In order to carry this out, we first of all transform the intervals $[t_1, 0]$ and $[t_2, 0]$ to $[-1, 0]$ by dividing through by $-t_i$, renaming the rescaled functions f_i'. We then form a convex combination:

$$pf_1' + (1 - p)f_2' : [0, 1] \to N_{location}$$

pointwise by taking:

$$(pf_1' + (1-p)f_2')(\tau) = pf_1'(\tau) + (1-p)f_2'(\tau) = pf_1((-t_1)\tau) + (1-p)f_2((-t_2)\tau)$$

Since both f_1 and f_2 are continuous in T, the result will be continuous in T.

The constraints on these sets are that $f_1(t_1)$ and $f_2(t_2)$ are in $C_{location}$ and that $f_1(0)$ and $f_2(0)$ are in $L_{location}$, and we need that their convex combinations are also in these respective locations. We know that $C_{location}$ and $L_{location}$ are convex, meaning that

$$pf_1(t_1) + (1 - p)f_2(t_2) \in C_{location} \text{ and } pf_1(0) + (1 - p)f_2(0) \in L_{location}$$

as required.

In this section, we have shown how we can use the interaction of words, represented by convex sets in conceptual spaces, to map sentence meanings down to a convex set in a conceptual space for sentences.

9.7 Discussion and Related Work

In this paper we have shown how grammar can be introduced to conceptual spaces theory by extending the categorical compositional scheme of Coecke et al. (2010). By using the kind of transformational compositionality that linguistic grammar introduces, we are able to extend compositional accounts beyond simply using variants of conjunction or disjunction that are seen in Aerts (2009), Hampton (1987, 1988a,b), and Lewis and Lawry (2016). This means that adjective-noun, verb-noun, and full sentence composition may be modelled, and the meanings of phrases and sentences are mapped into a shared meaning space.

In the current work, we have restricted ourselves to grammatical composition, and in particular pregroup grammar. However, the categorical compositional scheme can be instantiated in a number of ways. The grammar can be changed from pregroup grammar to another categorial grammar, as in Coecke et al. (2013a), or a compositional scheme that is not grammatically based may be used. Indeed, one of the challenges of this approach is to find a model of composition that accurately reflects human behaviour. One way of doing so would be to use an approach in which the syntactic scheme is generated by the semantics of the universe of discourse. Furthermore, since phrases and sentences are represented as sets equipped with a convex algebra, the model can in future work be extended to include logical composition.

9.7.1 Related Work

Our general aim of integrating compositional and semantic aspects of cognition is a longstanding problem in AI. In the Integrated Connectionist-Symbolic framework (Smolensky and Legendre 2006), the authors aim to integrate symbolic and connectionist reasoning, showing how symbolic reasoning can be instantiated in a neural network. One of the drawbacks of this approach is that sentence and phrase representations are dependent on the number of words in the phrase, so that it is difficult to directly compare noun phrases of different length, for example. That work inspired the development of the categorical compositional distributional semantic model of sentence meaning (Coecke et al. 2010) which the current work extends.

Our research also has links to cognitive architectures which integrate compositional and semantic aspects of cognition. Examples of these are Nengo (Eliasmith 2013), neural blackboard architectures (Van der Velde and De Kamps 2006), and the previously mentioned (Smolensky and Legendre 2006). Further, the use of conceptual spaces in cognitive architectures is an area of active research, as seen in, for example, Forth et al. (2016) and Lieto et al. (2017). Whilst the current work does not model key aspects of human cognition such as dynamics or action selection, it provides a model of compositionality that can be utilized by these architectures in representing inputs and how they may combine to form novel representations.

There is a wide range of research into modelling meaning and compositionality within conceptual spaces to which the theory developed here relates. Conceptual spaces have been formalized in Rickard et al. (2007), Adams and Raubal (2009), Lawry and Tang (2009), and recently in Bechberger and Kühnberger (2017). However the type of compositionality defined covers conjunction, disjunction, and correlations between domains, whereas the framework we propose covers a far more general type of compositionality. In Warglien et al. (2012), the authors develop a theory of verbs within the conceptual spaces framework. That work presents a model of *events* as consisting of at least a force vector and a result vector, together with a patient. It is also possible that an event has an agent, and an intention vector, amongst others. Verbs are then seen as being vectors that effect change, and which correspond only to one domain in the conceptual space. Our model has a similar goal in that it views a verb as a relation from one or more nouns to the sentence space. It therefore transforms nouns into sentences, which could be seen as events. However, rather than use a vector model of the verb, we use convex relations. The model that we propose seems to be more flexible than the two-vector approach advocated by Warglien et al. (2012). For example, consider the verb 'ripen'. In Warglien et al.'s approach, the effect of the verb on the noun is expressed as a vector moving the object representing the noun to another point in the space – called the result vector. However, the effect of 'ripen' on the noun 'banana' moves the points in the colour domain from green to yellow, whereas its effect on the noun 'plum' moves the points in the colour domain from green to purple. This can't be expressed using just one vector. The relational approach we use can take account of this however, by taking into account all the other dimensions in the description of the noun. Admittedly this is a more complex description, but arguably necessarily so. Derrac and Schockaert (2015) use a data-driven approach to developing semantic relations within conceptual spaces. The relations obtained are modelled as directions within the conceptual space. McGregor et al. (2016) develop a model of analogy within conceptual spaces, leveraging the geometrical properties of the meaning space.

9.8 Conclusion and Future Work

We have applied the categorical compositional scheme to cognition and conceptual spaces. In order to do this we introduced a new model for categorical compositional semantics, the category **ConvexRel** of convex algebras and binary relations respecting convex structure. We consider this model as a proof of concept that we can describe convex structures within our framework. Conceptual spaces are often considered to have further mathematical structure such as distance measures and notions of convergence or fixed points. It is also possible to vary the notion of convexity under consideration, for example by considering a binary betweenness relation on a space as primitive, rather than a mixing operation.

On the theoretical side, identifying a good setting for rich conceptual spaces models, and incorporating those structures into a compositional framework is a direction for further work, building on Marsden and Genovese (2017) and Coecke et al. (2017). Other theoretical work to be done includes investigating the implementation of logic within **ConvexRel**, specifically some form of negation, which in general does not preserve convexity.

On the more practical side, future work includes implementation within a data-derived conceptual space. This could include linguistic data derived from a corpus, but the generality of the conceptual spaces framework may also encompass a wider range of inputs such as from visual stimuli. Following on from this, future work will investigate integration with a cognitive architecture that encompasses dynamics and action.

Acknowledgements This work was partially funded by AFSOR grant "Algorithmic and Logical Aspects when Composing Meanings", the FQXi grant "Categorical Compositional Physics", and EPSRC PhD scholarships.

References

Adams, B., & Raubal, M. (2009). A metric conceptual space algebra. In *Spatial Information Theory* (pp. 51–68). Berlin/Heidelberg: Springer.

Aerts, D. (2009). Quantum structure in cognition. *Journal of Mathematical Psychology, 53*(5), 314–348.

Bankova, D., Coecke, B., Lewis, M., & Marsden, D. (2017). Graded entailment for compositional distributional semantics. *Journal of Language Modelling*, under review.

Bechberger, L., & Kühnberger, K. (2017). A thorough formalization of conceptual spaces. *CoRR*, abs/1706.06366.

Bolt, J., Coecke, B., Genovese, F., Lewis, M., Marsden, D., & Piedeleu, R. (2016). Interacting conceptual spaces. In D. Kartsaklis, M. Lewis, & L. Rimell (Eds.), *Proceedings of the 2016 Workshop on Semantic Spaces at the Intersection of NLP, Physics and Cognitive Science, SLPCS@QPL 2016*, Glasgow, Scotland, 11 June 2016 (EPTCS, Vol. 221, pp. 11–19).

Bullinaria, J., & Levy, J. (2007). Extracting semantic representations from word co-occurrence statistics: A computational study. *Behavior Research Methods, 39*(3), 510–526.

Carboni, A., & Walters, R. (1987). Cartesian bicategories I. *Journal of Pure and Applied Algebra, 49*(1), 11–32.

Coecke, B. (2013). An alternative gospel of structure: Order, composition, processes. In C. Heunen, M. Sadrzadeh, & E. Grefenstette (Eds.), *Quantum physics and linguistics. A compositional, diagrammatic discourse* (pp. 1–22). Oxford: Oxford University Press.

Coecke, B., & Kissinger, A. (2017). *Picturing quantum processes. A first course in quantum theory and diagrammatic reasoning*. Cambridge: Cambridge University Press.

Coecke, B., & Paquette, E. (2011). Categories for the practising physicist. In B. Coecke (Ed.), *New structures for physics* (pp. 173–286). Berlin/Heidelberg: Springer.

Coecke, B., Sadrzadeh, M., & Clark, S. (2010). Mathematical foundations for a compositional distributional model of meaning. *Linguistic Analysis, 36*, 45–384.

Coecke, B., Grefenstette, E., & Sadrzadeh, M. (2013a). Lambek vs. Lambek: Functorial vector space semantics and string diagrams for Lambek calculus. *Annals of Pure and Applied Logic, 164*(11), 1079–1100.

Coecke, B., Pavlović, D., & Vicary, J. (2013b). A new description of orthogonal bases. *Mathematical Structures in Computer Science, 23*, 555–567. arXiv:quant-ph/0810.1037.

Coecke, B., Genovese, F., Lewis, M., Marsden, D., & Toumi, A. (2017). Generalized relations for linguistics and cognition. *Theoretical Computer Science*, under review.

Derrac, J., & Schockaert, S. (2015). Inducing semantic relations from conceptual spaces: A data-driven approach to plausible reasoning. *Artificial Intelligence, 228*, 66–94.

Eliasmith, C. (2013). *How to build a brain: A neural architecture for biological cognition*. Oxford: Oxford University Press.

Forth, J., Agres, K., Purver, M., & Wiggins, G. (2016). Entraining IDyOT: Timing in the information dynamics of thinking. *Frontiers in Psychology, 7*, 1575.

Gärdenfors, P. (2004). *Conceptual spaces: The geometry of thought*. Cambridge/London: The MIT Press.

Gärdenfors, P. (2014). *The geometry of meaning: Semantics based on conceptual spaces*. Cambridge/MA: MIT Press.

Grefenstette, E., & Sadrzadeh, M. (2011). Experimental support for a categorical compositional distributional model of meaning. In *The 2014 Conference on Empirical Methods on Natural Language Processing* (pp. 1394–1404). arXiv:1106.4058.

Hampton, J. (1987). Inheritance of attributes in natural concept conjunctions. *Memory & Cognition, 15*(1), 55–71.

Hampton, J. (1988a). Disjunction of natural concepts. *Memory & Cognition, 16*(6), 579–591.
Hampton, J. (1988b). Overextension of conjunctive concepts: Evidence for a unitary model of concept typicality and class inclusion. *Journal of Experimental Psychology: Learning, Memory, and Cognition, 14*(1), 12.
Jacobs, B. (2011). Coalgebraic walks, in quantum and Turing computation. In *Foundations of Software Science and Computational Structures* (pp. 12–26). Berlin/Heidelberg: Springer.
Kamp, H., & Partee, B. (1995). Prototype theory and compositionality. *Cognition, 57*(2), 129–191.
Kartsaklis, D., & Sadrzadeh, M. (2013). Prior disambiguation of word tensors for constructing sentence vectors. In *The 2013 Conference on Empirical Methods on Natural Language Processing* (pp. 1590–1601). ACL.
Kartsaklis, D., Sadrzadeh, M., Pulman, S., & Coecke, B. (2013). Reasoning about meaning in natural language with compact closed categories and Frobenius algebras. In J. Chubb, A. Eskandarian, & V. Harizanov (Eds.), *Logic and algebraic structures in quantum computing* (pp. 199–222). Cambridge: Cambridge University Press (CUP).
Lambek, J. (1958). The mathematics of sentence structure. *American Mathematics Monthly, 65*, 154–170.
Lambek, J. (1999). Type grammar revisited. In *Logical aspects of computational linguistics* (pp. 1–27). Berlin/New York: Springer.
Lawry, J., & Tang, Y. (2009). Uncertainty modelling for vague concepts: A prototype theory approach. *Artificial Intelligence, 173*(18), 1539–1558.
Lewis, M., & Lawry, J. (2016). Hierarchical conceptual spaces for concept combination. *Artificial Intelligence, 237*, 204–227.
Lieto, A., Lebiere, C., & Oltramari, A. (2017). The knowledge level in cognitive architectures: Current limitations and possible developments. *Cognitive Systems Research, 48*, 39–55.
Lund, K., & Burgess, C. (1996). Producing high-dimensional semantic spaces from lexical co-occurrence. *Behavior Research Methods, Instruments, & Computers, 28*(2), 203–208.
Mac Lane, S. (1971). *Categories for the working mathematician*. New York: Springer.
Marsden, D., & Genovese, F. (2017). Custom hypergraph categories via generalized relations. In F. Bonchi & B. König (Eds.), *7th Conference on Algebra and Coalgebra in Computer Science, CALCO 2017*, Ljubljana, Slovenia, 12–16 June 2017 (LIPIcs, Vol. 72, pp. 17:1–17:16). Schloss Dagstuhl – Leibniz-Zentrum fuer Informatik.
McGregor, S., Purver, M., & Wiggins, G. (2016). Words, concepts, and the geometry of analogy. arXiv preprint arXiv:1608.01403.
Piedeleu, R., Kartsaklis, D., Coecke, B., & Sadrzadeh, M. (2015). Open system categorical quantum semantics in natural language processing. In L. S. Moss & P. Sobocinski (Eds.), *6th Conference on Algebra and Coalgebra in Computer Science, CALCO 2015* (LIPIcs, Vol. 35. pp. 270–289). Schloss Dagstuhl – Leibniz-Zentrum fuer Informatik.
Rickard, J., Aisbett, J., & Gibbon, G. (2007). Reformulation of the theory of conceptual spaces. *Information Sciences, 177*(21), 4539–4565.
Rosch, E., & Mervis, C. (1975). Family resemblances: Studies in the internal structure of categories. *Cognitive Psychology, 7*(4), 573–605.
Sadrzadeh, M. (2007). High level quantum structures in linguistics and multi agent systems. In *AAI Spring Symposium: Quantum Interaction* (pp. 9–16).
Sadrzadeh, M., Clark, S., & Coecke, B. (2013). The Frobenius anatomy of word meanings I: Subject and object relative pronouns. *Journal of Logic and Computation, 23*, 1293–1317. arXiv:1404.5278.
Sadrzadeh, M., Clark, S., & Coecke, B. (2016). The Frobenius anatomy of word meanings II: Possessive relative pronouns. *Journal of Logic and Computation, 26*(2), 785–815.
Smolensky, P., & Legendre, G. (2006). *The harmonic mind: From neural computation to optimality-theoretic grammar* (Cognitive architecture, Vol. 1). Cambridge: MIT Press.
Tversky, A. (1977). Features of similarity. *Psychological Review, 84*(4), 327–352.
Van der Velde, F., & De Kamps, M. (2006). Neural blackboard architectures of combinatorial structures in cognition. *Behavioral and Brain Sciences, 29*(1), 37–70.
Warglien, M., Gardenfors, P., & Westera, M. (2012). Event structure, conceptual spaces and the semantics of verbs. *Theoretical Linguistics, 38*(3–4), 159–193.

Chapter 10
Magnitude and Number Sensitivity of the Approximate Number System in Conceptual Spaces

Aleksander Gemel and Paula Quinon

Abstract In this paper, we propose a conceptual-spaces model of numerical cognition, and more precisely, of representations generated by Approximate Number System. The model is an extended and improved version of our earlier result (Gemel A, Quinon P: The approximate numbers system and the treatment of vagueness in conceptual spaces. In: Lukowski L, Gemel A, Zukowski B (eds) Cognition, meaning and action. Jagiellonian-Lodz University Press, Kraków, pp 87–108, 2015), where only purely quantitative information was accounted for. We focused on the idea that ANS evolved to detect numerosity in the input, and to abstract this information from all possible magnitude-related cues, such as size of compared objects, aggregate area of those objects, or density of their location. The idea is that when one sees a pile of apples, ANS acts as a "number sense", informing one of the approximate quantity of apples. Consequently, the original model was very simplified, accounting only for one, uniform perceptual discrete visual input. With inspiration from computational models (ex. Dehaene S, Changeux JP: J Cogn Neurosci 5:390–407, 1993;, Lourenco SF, Longo MR: Psychol Sci 21(6):873–881, 2010) that can process more complex stimuli, we propose in this paper a conceptual-spaces model for non-uniform input. The improved version of the model accounts additionally for magnitude-related cues related to both "number sense" and "magnitude sense".

A. Gemel (✉)
Department of Cognitive Science, Institute of Psychology, University of Lodz, Lodz, Poland
e-mail: aleksander.gemel@gmail.com

P. Quinon (✉)
Department of Administration and Social Sciences, Warsaw University of Technology, Warsaw, Poland

Department of Philosophy and Cognitive Sciences, Lund University, Lund, Sweden
e-mail: paula.quinon@fil.lu.se

M. Kaipainen et al. (eds.), *Conceptual Spaces: Elaborations and Applications*, Synthese Library 405, https://doi.org/10.1007/978-3-030-12800-5_10

10.1 Introduction

Since the early 1960s, researchers in numerical cognition have postulated the existence of *a priori* intuitions that precede – and then structure – how both humans and many non-human animals experience and process quantitative information. Two programs are particularly well-explored. First, Dehaene and Brannon (2010) offer a "Kantian research program" building explicitly on a neo-Kantian philosophical framework. Second, Elisabeth Spelke (2000) writes of "core cognition": a concept that Susan Carey (2009) has largely explored in the context of numerosity. With minor differences, both research programs postulate the existence of innate cognitive mechanisms responsible for processing of quantitative information – even as the nature of this information is still very much the subject of debate in number-cognition studies. The Approximate Number System (ANS) is the innate core cognitive systems, described by Dehaene (1997/2011), that provokes particularly many discussions. The main controversy concerns the content of representations generated by the ANS. Our objective in this paper is to present a coherent model of the quantitative representations employed in the ANS, in pursuit of a better understanding of the correlations between number and magnitude sense.

Leibovich et al. (2017) argue that the system processing quantitative information – at least at the earliest stages of human development – does not address numerosity of discrete entities, as the Approximate Number System does (Dehaene 1997/2011), but only the magnitude of the space they cover. Leibovitz's argument builds on the idea that the perception of infants – though able to process quantitative information – is not acute enough to distinguish small, individual objects (Banks 1980; Dobson and Teller 1978). Carey (2001; see also Feigenson et al. 2002a, b) writes that infants younger than 5 months are not able to individuate objects and claims that such individuation is necessary for all numerical quantity assessment. In consequence, humans must be equipped with some Analogue Magnitude Representations (AMRs) system: e.g., the system taking as input a surface covered by discrete objects and returning as output an analogue representation of their spatial magnitude. This system might – but need not – be replaced or supplemented by the ANS that processes numerosity later on in child development.

It is complicated indeed to extract any strictly numerical aspect to the AMRs system's processing of quantities when activated by a non-symbolic input: e.g., a collection of dots distributed on a surface. Numerosity here strongly correlates with other dimensions, such as surface area, contour length, density, etc. Starkey and Cooper (1980: 1033) explain the problem this way:

> The bases of discriminations among large values are not known and may vary with different types of arrays (for example, dot size, the spatial, distance separating adjacent dots, or surface area of the total array). The ability underlying these discriminations might be the same as that which allows adults to estimate the more numerous of two arrays differing substantially in number.

Mix et al. (2002a: 293) claim that the processing of numerosity by ANS is just a byproduct of the processing of continuous input:

> Contrary to what has been claimed in current models of early quantification, we failed to find evidence that discrete number is represented in infancy. Instead, there is strong evidence that infants view spatial quantities in terms of total amount of substance. There are also indications that infants view sequential quantities in terms of temporal cues, such as rate, duration, and rhythm.

In this paper we will amount to this discussion by proposing a model of numerical cognition. Cognitive modelling discloses coherence and enhances explanatory power of experimental investigation. We will expand the conceptual-spaces model of the ANS proposed by Gemel and Quinon (2015). The original model is very simplified, accounting only for one, uniform perceptual discrete numerical input (Fig. 10.1). With inspiration from computational models such as those of Dehaene and Changeux (1993) and Lourenco and Longo (2010) that use more complex stimuli, we will propose in this paper a conceptual spaces model for non-uniform input (Fig. 10.2).

Given its simplicity, our earlier model cannot distinguish reaction to numerosity from the space occupied by discrete elements. In consequence, it does not provide any new input to discussion of AMRs systems and the way they process quantities.

Fig. 10.1 Uniform perceptual input for the model proposed by Gemel and Quinon (2015)

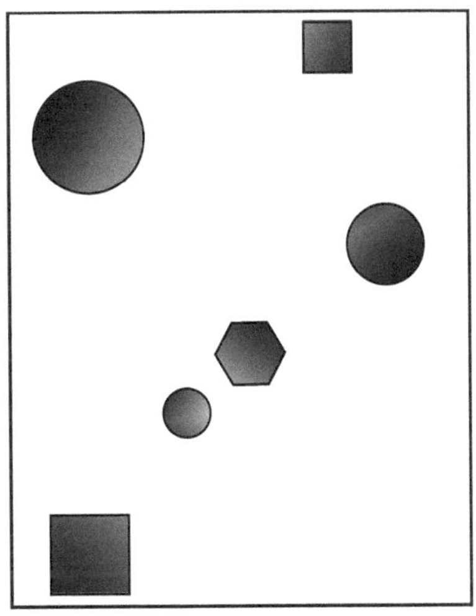

Fig. 10.2 Non-uniform, non-symbolic perceptual input

Our new model allows us to add a mechanism for representing the numerosity of non-symbolic stimuli – by constructing a conceptual space of ANS representations, generated in response to input of randomly distributed discrete entities of varying size and shape (Fig. 10.2).

Modelling non-uniform input necessitates an explicit account of the system responsible for processing of quantities.

10.2 Numerosity vs. Magnitude

As said, abundant evidence exists that preverbal children process quantitative information; Xu and Spelke (2000; see also Xu 2000) provide evidence for children as young as 6 months. The system responsible for this processing displays characteristics of an AMRs system. Most importantly, it is subject to Weber's law, according to which successful comparison of two quantitative inputs does not depend on the absolute difference between sizes but depends on the ratio. We know that infants apprehend the difference between sets that sizes are in the ratio 1:2 (e.g., eight vs. sixteen elements) but not the difference between sets in the ratio 2:3 (Xu and Spelke 2000). We also know that acuity in comparison increases over development; adults are able to apprehend correctly ratios of 9:10 (Halberda and Feigenson 2008).

Systems subject to Weber's law, including an AMRs system processing quantities, are subject to two consequent effects. The *magnitude effect* says that assessment is easier for smaller magnitudes; in the case of quantities, it is easier to compare ten with twenty elements than 200 with 210. According to the *distance effect* the larger the ratio, the easier to assess the difference: it is easier to compare ten with twenty elements than ten with eleven. Any model of AMRs systems needs to take these effects into account.

Traditionally, the AMRs system that processes quantities is identified with the Approximate Number System. It is thought to process numerical information (Libertus et al. 2011):

> One central component of the number sense is the Approximate Number System (ANS). A focus of research in cognitive psychology and neuroscience, the ANS has been shown to support a primitive sense of number in infants, children, and adults.... The ANS has been shown to produce imperfect "noisy" estimates of numbers of items from input across all sensory modalities (e.g., beeps, visually or tactilely presented objects, taps of a finger). These numerical estimates support quantitative computations such as "greater-than, less-than," addition, subtraction, multiplication, and division.

Recent research suggests that infants have the capacity to distinguish sets of discrete elements on the basis of e.g. surface occupied, volume, density, and medium radius of the space (Gebuis and Reynvoet 2012a; see also Gebius and Reynvoet 2012b):

> In numerical cognition, the theory of the approximate number system predominates research about our ability to compare or estimate number. The theory suggests that this system

> enables us to extract number independently of the continuous visual variables that are present in nonsymbolic number stimuli (e.g., the diameter or convex hull of an array of dots).... As an alternative to the approximate number system, we previously proposed that nonsymbolic number judgments are based on the different visual properties comprising the number stimuli.

Experimental results confirm that numerosity is not the only feature that can be used to generate representations of sets of discrete elements through an AMRs system. Leibovich et al. (2017) – among others – claim that numerosity appears at later stages of development; small children assess quantity on the basis of spatial extension. Feigensen et al. (2002a, b) observed 10- and 12-month infants choose a plate of crackers with overall greater volume over one with greater numerosity. Slusser and Sarnecka (2011) tested 116 thirty- and 48-month infants on a task where they were shown a picture of a herd of big turtles. They were then asked to decide which picture it matched: one with the same number of small turtles or one with twice as many small turtles covering the same surface area as in the original picture. Only children with experience of counting, who understood the names of numbers, could correctly perform the task.

Mix and colleagues (2002b) take a similar view. Their thesis is based on the assumption that infants both discriminate and quantify sets by using non-numerical perceptual cues such as area, volume, and length. They reject a "double nature" of the ANS, claiming that all processing of non-symbolic, quantitative information takes continuous values as input. They write (2002b: 62), "development starts with only one principle of quantification in infancy based on amount of substance, which applies to both continuous quantity and sets of discrete objects." They continue (2002b: 81), "infants do not represent quantities numerically at all. Instead, the evidence points to the use of overall amount."

Observe, that the discussion concerns a nature of representations generated by the ANS. Proponents of both theories agree that the input contains both, discrete (numerical) and continuous information. It is also commonly assumed that there exists an AMRs system that transforms exact information of the input into an approximate representation. What remains unclear is the character of the main conceptual content represented in an output: numerosity or magnitude (Fig. 10.3).[1]

Dehaene (1997) opts for numerosity, claiming that discrete representations are generated from the discrete input by the ANS. The ANS evolved to detect numerosity in the input and abstract from its magnitude-related cues such as size of compared objects, aggregate area of those objects, or density of their location. When one sees a pile of apples; the ANS acts as a "number sense", informing one of the approximate quantity of apples. Nieder and Dehaene (2009) see this process being as basic and automatic as perceiving the perceptual features of objects. Similarly, Burr and Ross (2008: 425) write:

[1] In certain models it is difficult to distinguish the main conceptual content and one could claim that the two are conveyed in an equal manner. Getting into detailed discussions is beyond our focus in this paper.

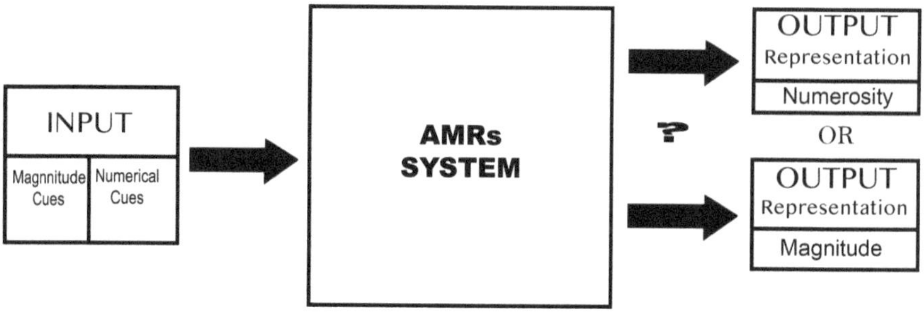

Fig. 10.3 Simplified scheme of the possible representational structure of an AMR system

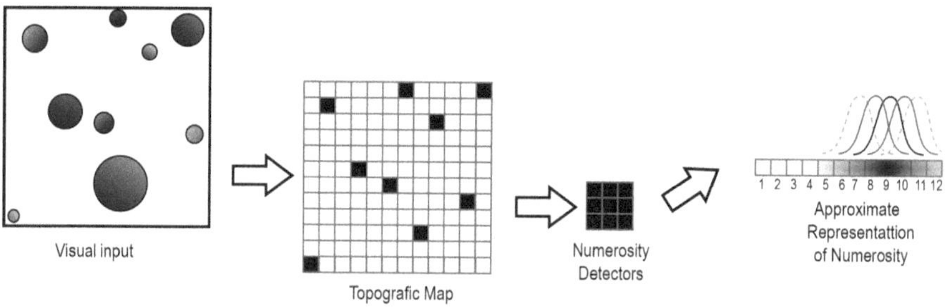

Fig. 10.4 Connectionist model of ANS proposed by Dehaene and Changeux (1993)

> Perceived numerosity is susceptible to adaptation, like primary visual properties of a scene, such as color, contrast, size, and speed. Apparent numerosity was decreased by adaptation to large numbers of dots and increased by adaptation to small numbers, the effect depending entirely on the numerosity of the adaptor, not on contrast, size, orientation, or pixel density, and occurring with very low adaptor contrasts. We suggest that the visual system has the capacity to estimate numerosity and that it is an independent primary visual property, not reducible to others like spatial frequency or density of texture.

Dehaene and Changeux (1993) propose a computational model where the numerical aspect of the output representation is particularly clear. Input is taken to be non-symbolic, possibly non-uniform, visual stimuli – such as a group of objects (Fig. 10.4). The output is an approximate representation of the numerosity of elements. Their model is based on connectionist approach: i.e., modelling by artificial neural networks (ANNs).

Processing runs through several phases. In the first phase, objects are encoded according to location in space, as determined by information from the retina – producing a topographical map of object locations; this is represented in the ANN by activation of neurones. Abstraction is made from all the features related to continuous magnitudes, such as size of items and density of the set.

In the second phase, the numerosity of items represented in the first phase is estimated by numerosity detectors – specialized groups of neurons (Dehaene and Changeux 1993: 394) – reflecting the assumption that the numerosity-detecting capacity is innate. The role of the detectors is to sum the aggregate activation of active neuronal groups from the previous phase.

In the final phase, the neural network generates an approximate representation of the input's numerosity.

Verguts and Fias (2004) tested the model. After training the network, they showed that the distribution of errors in both discrimination and comparison tasks followed empirical data regarding two ubiquitous effects in numerical cognition: size and distance effects.

According to the defenders of "continuous magnitudes" reading, there exists no such system as the ANS that processes numerical information and generates continuous representations encoding the numerosity of discrete elements in a collection. "Number sense", if there is such a thing, develops later on ontogenetically, arising from systems that process continuous magnitudes. Instead, there exists an innate "sense of magnitude" enabling humans (and also many animals) to differentiate between continuous magnitudes, *i.e.*, the whole surfaces that the assessed items cover (Gebuis and Reynvoet 2012b; Leibovich and Ansari 2016; Leibovich et al. 2015; Mix et al. 2002a, b).

In theories exploring the "sense of magnitude", the processing of non-symbolic, quantitative input is holistic. Numerosity is not extracted from non-symbolic input independent of information about surface magnitude: size of object stimulus, aggregate area of stimulus, stimulus density, etc. All this information is processed simultaneously, so that the representations generated have a double nature. Leibovich and Henik (2013: 2) write:

> We argue here that since it is physically impossible to study discrete magnitudes in isolation, it is difficult to accept theories such as the ANS and the number senses, which suggest numerosity is the most salient cue. It is more likely that we integrate multiple visual cues to make size estimations.... Following this logic, it is more instrumental to use all available cues in the environment – not only discrete ones – to make size estimations.

Drawing on the idea of a natural correlation between continuous magnitudes and discrete magnitudes (numerosity), where all available types of information are used to judge numerosity, Leibovich and Henik propose a model for an AMRs system within a developmental framework, describing how magnitudes are represented/discriminated at different developmental stages (Fig. 10.5):

According to Leibovich and Henik, humans are born with only the ability to discriminate continuous properties. At this first stage of development, infants must judge stimuli by relying solely on continuous magnitude; their vision is not acute enough to focus on specific items. It is later during development – after infants reach the age of 5 months and are finally able to individuate items from background and one another – that they represent both continuous and discrete properties and start to learn the correlations between the two. When the process is complete, babies are able to use both discrete and continuous properties to estimate magnitude. At the final stage of development, upon starting formal math education, children are able to represent the exact difference between magnitudes. However, even after formal training, the representation of non-symbolic stimuli as processed by the

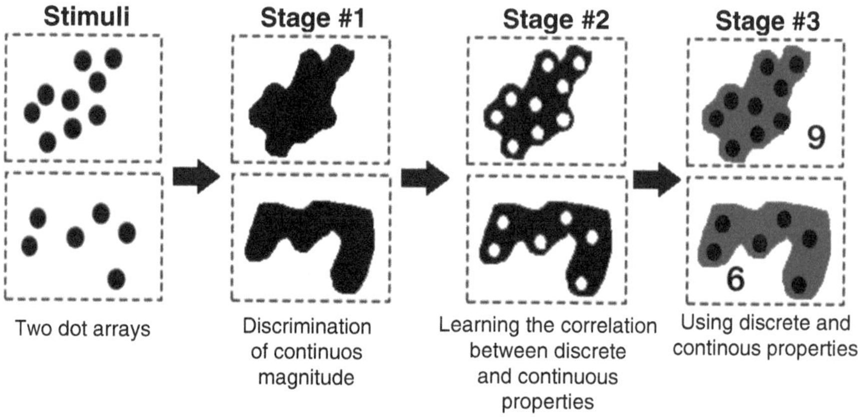

Fig. 10.5 Developmental model proposed by Leibovich and Henik (2013)

AMRs system is holistic, taking into account both discrete and continuous aspects of input. The outputs are representations that convey the information about the discrete quantity of elements in the assessed collection entwined in the information about the surface they cover.

Neither interpretation – discrete vs. continuous magnitude – has gathered sufficient empirical support to exclude the other fully. Each underlying theoretical framework – connectionist, developmental, conceptual-spaces based – necessitates deciding which type of input – discrete or continuous – ultimately prevails. A robust, consistent model would help empirical researchers formulate hypotheses in one or the other direction. We propose a model where output representations derive from discrete input.

We are able to improve upon our earlier results (Gemel and Quinon 2015) by showing how number-related cues can be represented in an ANS model structured within conceptual spaces. In the earlier model, the output representations contain purely quantitative information; but their presentation offers no way to distinguish numerosity-related from magnitude-related cues. Differentiation is not possible because the model is defined by one type of uniform perceptual input only (Fig. 10.6a), where the number of elements and surface area occupied change proportionally. We extend that model, using a conceptual space to depict the conceptual structure of the ANS-generated numerical information, working with non-symbolic, non-uniform input (Fig. 10.6b).

Our objective, again, is to present a coherent model of the quantitative representations employed in an ANS, in pursuit of a better understanding of the correlations between number and magnitude sense. We set aside any other general questions relating to the cognitive processing of quantitative information. We make no attempt to determine how number and magnitude sense are correlated, nor which theory is ultimately correct. Our modest aim is simply a better understanding of the underlying processes.

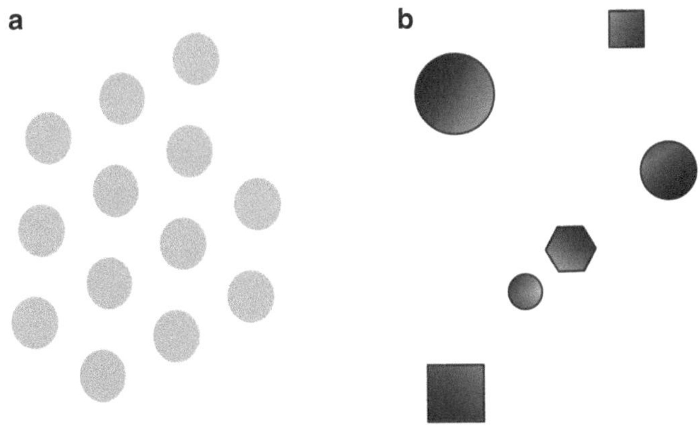

Fig. 10.6 (**a**) Non-symbolic, uniform perceptual input (**b**) Non-symbolic, non-uniform perceptual input

We extend our earlier model by defining explicitly a number-specific module sensitive to stimulus numerosity, inspired by Dehaene and Changeux's (1993) aforementioned numerosity detectors. Modelling in conceptual spaces lets us avoid major problems with their connectionist model. First of all, it avoids the necessity of providing a (large) training set for developing the requisite structures. The reason why Dehaene and Changeux's model learns so slowly is that it has so many learning parameters: all the weights of all the neuronal connections are treated as independent variables, and all must be adjusted over the course of training (see Gärdenfors 2000). This is the main reason their model is limited to five elements (Dehaene and Changeux 1993: 396).

Second, it is very difficult to express conceptual change in connectionist models: reconfiguration of the network is difficult and time consuming. Connectionist models are consequently awkward to use in any developmental framework: e.g., in one instantiating a theory of numerosity development.

Third, ANNs pose a fundamental epistemological problem because their representations are necessarily emergent. They offer no knowledge of representational content, only of representational implementation. Even if one has reason to believe that a network categorizes inputs appropriately, one cannot say just what it represents. All these difficulties may be avoided by modelling at a higher – conceptual – level of representation such as offered by conceptual spaces.

10.3 Modelling a Number-Sensitive Module in Conceptual Spaces

As noted, our earlier model could not deal with non-uniform input: i.e., elements not all of similar size and distributed in a regular manner. The addition of a number-sensitive module permits the transformation of discrete non-uniform input into continuous numerosity representation, with no need to represent brain-level computation – only to simulate those (higher) cognitive functions associated with ANS-generated representations. That said, we suppose that our model can be brought into alignment with Dehaene and Chagneux.

Our model, like theirs, consists of three phases. First, magnitude-related cues are abstracted from input stimulus (Sect. 3.1). Next, the numerical outputs of the first phase are encoded as continuous (analogue) representations (Sect. 3.2). Finally, the analogue outputs of the second phase are aligned with the prototype structure of the category (Sect. 3.3).

10.3.1 First Phase: Topographical Representation

Visual input is represented by a group of points in two-dimensional similarity space with quality dimensions height and length, encoding the spatial location of stimulus objects (Fig. 10.1). Distance in the similarity space is defined by a city-block metric:

$$d_{Cb}(x, y) = \sum_{k=1}^{n} | x_k - y_k |$$

Where $d_{Cb}(x,y)$ is the distance between points x and y, x_k and y_k their coordinates along quality dimension axes k, and n the number of quality dimensions constituting the space (in this case, two). Less formally, a city-block metric measures the distance between two points as the sum of the absolute values of the differences between their coordinates. It says how many steps one must take along the height dimension and how many along the width dimension (always keeping to the grid!) to get from Point x to Point y. Degrees on the scale in the proposed metric are defined by the relations of "being-above" and "being-below" for the height dimension, and by the relations of "being-to- the-left" and "being-to-the-right" for the length dimension. If y is one degree (one step) above x along the vertical axis, then y is directly above x and x directly below y; two degrees mean that there is another object between them. The same applies to relations of left and right. No assumption is made that the stimuli have an unambiguous centre point. In fact, such a point is not needed to order the objects in the grid defined by above, below, left, and right relations since one can do such ordering by starting from any object in visual stimulus.

The relation of above is defined as follows:

above:

$$\text{(Ax.1)} \quad \forall_x \forall_y xAy \Rightarrow \neg(yAx)$$

$$\text{(Ax.2)} \quad \forall_x xAx \Rightarrow x \in \varnothing$$

$$\text{(Ax.3)} \quad \forall_x \forall_y \forall_z xAy \wedge yAz \Rightarrow xAz$$

The "below" relation is dual to "above": i.e., $xBy \iff yAx$. Similarly, we can define the relations of left and right.

The "above" relation is gradual, allowing the additional relation "one degree above" (A_{1°) defined as:

$$\forall_x \forall_y \; x A_{1^\circ} y \iff x A y \wedge \neg \exists_z \; z A y \wedge z B x,$$

Similarly, we define the relations "one degree left" (L_{1°) and "one degree right" (R_{1°).

A_{1° and B_{1° define degrees of similarity on the height dimension, L_{1° and R_{1° degrees of similarity on the length dimension. As should be clear, the scale of the space is a normalized measure of the distance between x and y, taking into account only the number of steps between them along a grid and not the actual distance.

Representation of objects in this conceptual space allows abstracting away most of their magnitude-related cues: i.e.. size, shape, overall area occupied, etc. Each object is represented as a point in space. Use of a city-block metric allows abstracting away most information about physical distances between stimulus objects. The result is a structural map of direct "neighborhood" relations between stimulus objects (Fig. 10.7), defined as:

$$\forall_x \forall_y \; x DN y \iff \left(x A_{1^\circ} y \veebar x B_{1^\circ} y\right) \vee \left(x L_{1^\circ} y \veebar x R_{1^\circ} y\right)$$

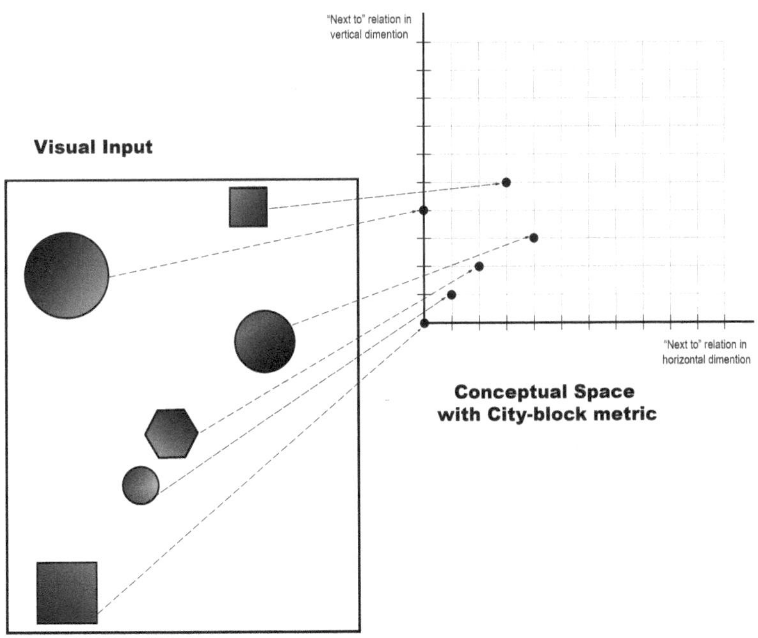

Fig. 10.7 First phase of heterogenic stimulus representation in conceptual space. Note that the topological structure of the stimulus is preserved

The formula says that two points are direct neighbors if they are in any of the simple one-degree relations (left, right, above, below) or in any one-degree-diagonal relation: i.e., one of the conjunctions A_{1° and L_{1°; A_{1° and R_{1°;B_{1° and L_{1°; or B_{1° and R_{1°.

10.3.2 Second Phase: Analogue Magnitude Representation of Numerosity

Next comes the encoding of numerical information into continuous (analogue) representations. Carey and Sarnecka write (2006, p. 477): "[there is] a helpful analogy to the following external system of analogue number representations. ...Line length is a direct analogue of number." (See Fig. 10.8).

The term "analogue" should be understood in accordance with Goodman (1976) as being *continuous*, or *dense*: between any two values represented values a third can always be identified. We believe that the best tools for describing analogue representations are spatial categories with spatial (quality) dimensions, where size (in this case line length) is proportional to quantity of elements (numerosity): the greater the quantity in the input, the larger the magnitude in the output representation.

The second phase represents the output of the first phase (topographical representation of a normalized set of points) as a polygonal chain running through each of the points. There are two sub-steps. First, the topographical representation is scaled according to the Euclidean metric defined on the similarity space as:

$$d_E(x, y) = \sqrt{\sum_{k=1}^{n} (x_k - y_k)^2}$$

Where the distance between any two points *x*,*y* is the length of the segment $\overline{xy}$ connecting them.

Next, we assume a simple algorithm that generates the polygonal chain by a clockwise ordering of *h* points $L_0 \ldots L_{h-1}$. The first point L_0 is typically the leftmost in set *P*. If multiple points in *P* have the same *x* coordinate, L_0 is the one with the smallest *y* coordinate. One can start though (step $i = 0$) from any point.

Fig. 10.8 Analogue representation of numerosity of non-symbolic stimulus

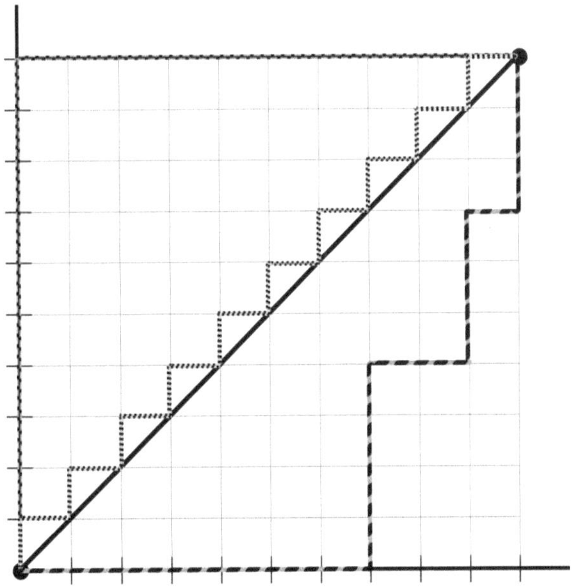

Fig. 10.9 The shortest distance according to the Euclidean metric (black solid line) has several equivalent distances according to city-block metric (gray dotted lines)

The next point L_{i+1} is the one with the smallest distance to L_0. Measurement of distance between points is carried out in accordance with the Euclidean metric, which (unlike the city-block metric) guarantees that the shortest route connects the points: e.g., there is only one way to connect two points lying diagonally (Fig. 10.9).

In case more than one point is equidistant to L_0 – e.g., L_0 has coordinate (1,1), L_i (1,2), and L_j (2,1) – then L_1 is the point with lowest value on the ordinate axis. This simple process is iterated as $i := i + 1$ over h steps until the end point is reached: *i.e.*, $L_h := L_h + 1 = L_0$. The last step is thus the one preceding a return to starting point L_0.

Of course, the algorithm's efficiency (including processing time) should be checked experimentally. It seems, that the topographical representation of a normalized set of points guarantees that computational complexity will be small enough to comply with human cognitive resources and, indeed, rules of cognitive economy. At the same time, we are aware that using a sequential algorithm to generate the polygonal chain increases the computational complexity with every increment in the number of objects in the stimulus: more objects means more steps. This effect is expected, given that estimating a larger number of objects usually takes more time. On the other hand, some empirical results from the ANS comparisons suggest that processing time is almost completely unrelated to number of stimulus objects. Dealing with this apparent problem requires an algorithm based on parallel instead of serial summation. A parallel algorithm generates a polygonal chain by connecting all points simultaneously: i.e., it starts from all points in P simultaneously, connecting any point a with nearest point b not lying to the left of a. If more than one point matches, then b is the point with lowest value along

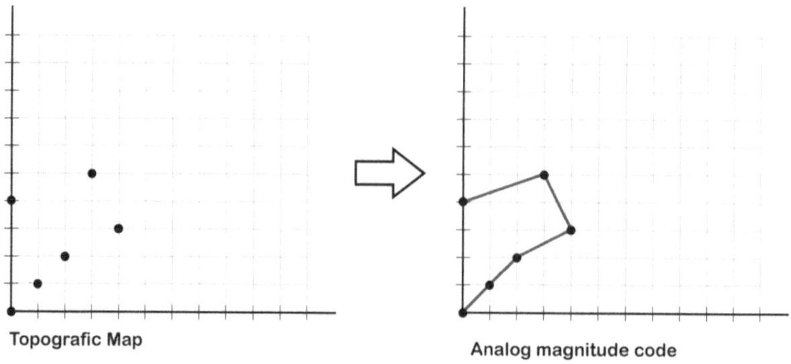

Fig. 10.10 Transformation of topographic map into continuous (analogue) representation

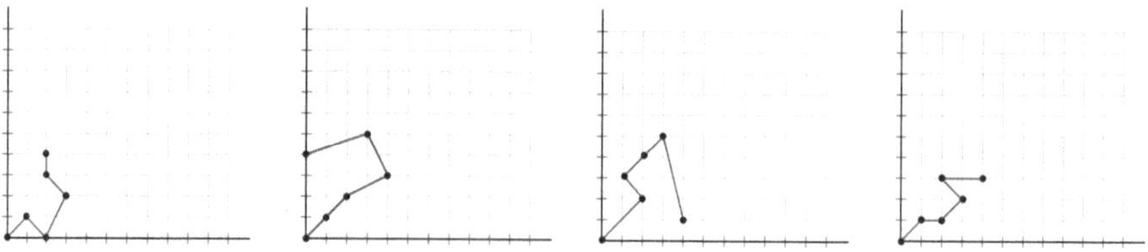

Fig. 10.11 Possible AMRs with equinumerous input

the x-axis. The operational time of the algorithm is reduced to just a few steps, regardless of the number of objects.

Regardless of which algorithm is used, a collection of discrete points from the first phase has been transformed into a continuous (analogue) representation (Fig. 10.10).

The numerosity of a given representation – the number of elements on which the algorithm operates – is, again, directly proportional to the number of steps generating the polygonal chain. Even if the continuous representations generated by our model might differ significantly even between equinumerous inputs (Fig. 10.11), each will eventually have the same numerical dimension, corresponding to the number of steps.

Numerosity as a qualitative dimension is represented by a third axis z, isomorphic to the positive numbers: i.e., minus the negative numbers and zero. (Negative and zero numerosity cannot be represented through non-symbolic stimuli.) Observe that it only represents the numerosity of the stimulus when the stimulus consists of more than one element: an ANS affords no possibility of an approximate representation of individuality (Feigenson et al. 2004; Xu 2003). Note, too, that the numerosity of the stimulus is always one degree greater than the number of steps taken: i.e., one degree along the numerical dimension means that the algorithm has taken *one* step to connect *two* points (i.e., two objects). One step is represented as a continuous line one-degree long; since the scale starts with one, the line will have a length of

Fig. 10.12 The numerical dimension. *A* represents Step 0 of the polygonal chain algorithm, corresponding to a numerosity of approximate one or zero elements. *B* represents Step 1, corresponding to a numerosity of approximate two, etc.

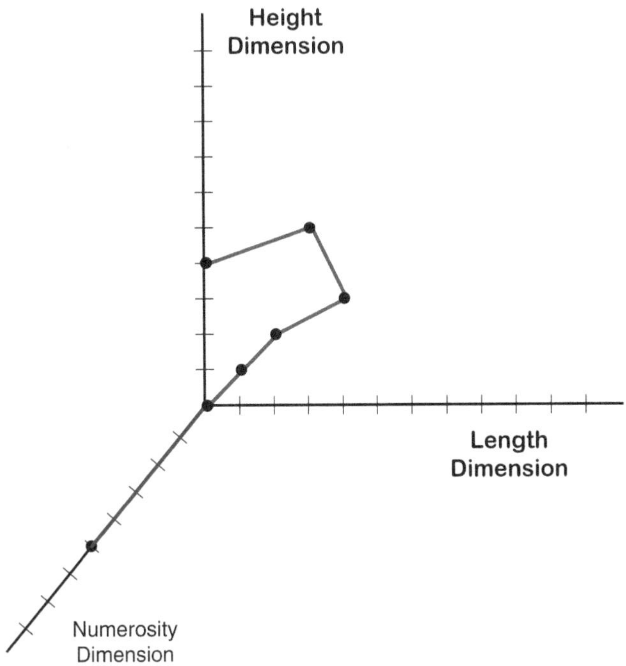

Fig. 10.13 Second phase output: an analogue magnitude representation of numerosity

about two degrees, corresponding to the numerosity of the stimulus. In general, the numerical value of an n-element stimulus will be represented as a line of length $i = n - 1$ (Fig. 10.12).

We assume that, in the analogy to the solution proposed by Dehaene & Changeux, numerical representation is generated as a result of summation. In our proposal, the algorithm generating polygonal chain sums up number of its own steps, and represents them in analogue form i.e. as the length of a line in another quality dimension. The analogue nature of the representation means that the ANS cannot represent an exact number (indeed, the representations are not only non-discrete but also noisy); but that is to be expected. Were there such a capacity, the brain would need to do additional processing to come up with an imprecise estimation – and what would be the point?

The output of the second phase is a three-dimensional similarity space of dimensions length, height, and numerosity (Fig. 10.13).

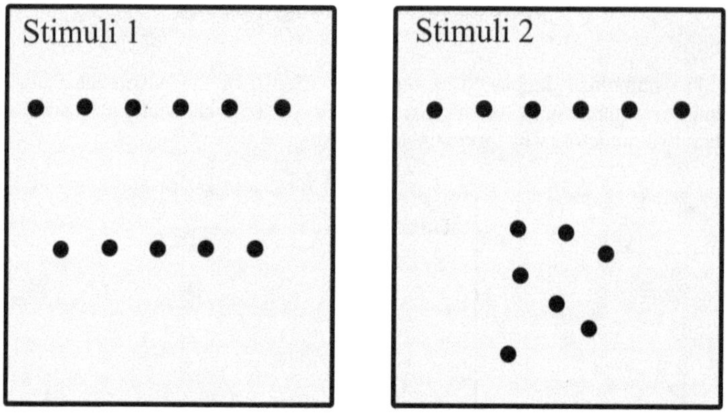

Fig. 10.14 Differences in difficulty distinguishing numerosity, given different stimulus arrangement

Note that, while different stimuli may have the same numerosity, some will be more recognizable than others: i.e., the cognitive effort needed to distinguish the numerosity of two stimuli depends not only on the number of objects, but also on their arrangement. Comparing the numerosity of two linearly arranged stimuli (Stimuli 1, Fig. 10.14) is easier (involves less cognitive effort) than distinguishing that of a linearly arranged stimulus from one arranged randomly (Stimuli 2, Fig. 10.14).

Our model offers an intuitive explanation via a mechanism for ordering ANS representations according to degree of complexity of stimulus object arrangement, based on the prototype structure of the representation. According to Prototype Theory, the internal structure of a category is not homogeneous: i.e., there may be different degrees of representativeness within it; some elements (the prototypes) are more centrally located within categories than others. Elements more similar to the prototype are situated closer to the centre and characterized by a greater degree of membership; those less similar have a more peripheral location.

10.3.3 Third Phase: Prototype Representation of ANS Representations

The third phase organizes representations according to their similarity to the relevant prototypes. Certain distributions of the kind of visual input modeled in our study we take to be cognitively privileged. As *per* our earlier proposal (Gemel and Quinon 2015), we justify the existence of prototypical quantitative representations based on psychological investigations.

The first and second phases of our model preserve the topological structure of the stimulus (Fig. 10.7) – enriched, in the second phase, with the quality dimension of numerosity (Fig. 10.13). Again, different stimuli with the same numerosity can have completely different topologies (Fig. 10.11), with direct impact on cognitive effort.

We use topographical properties of the stimulus to define the prototype structure of ANS-generated representations, where estimation of numerosity for prototypes will be much easier than for peripheral elements – yielding more incorrect estimates for the latter. Putting this another way, some arrangements of elements in the stimulus, as represented by the polygonal chain, will be more easily recognized as corresponding to a given numerosity and so involve less cognitive effort and less processing time.

Our earlier proposal (Gemel and Quinon 2015) was that the prototypical arrangement is the most evenly distributed one – i.e., a square arrangement of equal height and length – justified on the basis of Tversky's (1977) studies on the prominence of geometric figures. According to Tversky, the more regular and symmetrical the shape of a figure, the more prominent (salient) it becomes. However, empirical research should be conducted to determine the actual prototypical arrangements. Until such a study is conducted, we do not want to prejudge the outcome.

For sake of simplicity, we propose that prototypes are represented by the straight-line arrangement of objects. However, we cannot exclude the hypothesis that dots distributed in certain ways are more prototypical than others or that more than one shape is prototypical. Obviously, such cases would significantly affect similarity relationships within the conceptual space and consequently the final shape of the ANS-generated representation.

The hypotheses we have presented could be tested empirically through a series of categorization experiments. Our ambition is to proceed to preparing the necessary theoretical foundations for such studies.

Again, the third phase of our model orders all possible configurations of continuous (analogue) representations associated with a given numerosity according to their similarity to prototypical elements (Fig. 10.15). The ordering is made in purely spatial terms: i.e., by similarity in height and width dimensions of the polygonal chains generated in the second phase.

By introducing prototypes, we create a suitable structure for all ANS-generated representations, locating them within one common space that takes the form of a Voronoi tessellation – as proposed by Gärdenfors (2000). Formally, the tessellation is expressed by the formula:

$$VorP\left(p_{1}\right)=\left\{x \in E \mid \forall p_{n} \in S, d\left(x, p_{1}\right) \leq d\left(x, p_{n}\right)\right\}$$

Where $S = \{p_1, \ldots, p_n\}$ is the set of prototypes, E the Euclidean-metric space, and d a distance function on E correlated to similarity between stimuli. If one starts from set S of prototypes represented by points (p_1, ..., p_n) and assumes that E is a Euclidean-metric space, then – for every point x in that space – one can measure the

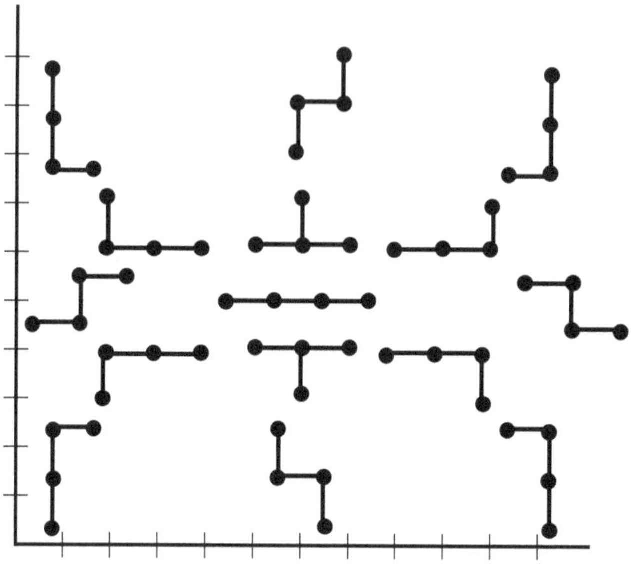

Fig. 10.15 Ordered set of arrangements of four-element collections of ANS stimuli

distance to every element of the set. The space is then partitioned according to the rule that each *x* belongs to the same region *P* as the closest prototype from set *S*.

In our model, we have assumed that more than one prototype will be assigned to any given numerosity. If we stipulate that prototypicality is based on linear shape, then the possible linear arrangements of stimulus elements are horizontal, vertical, and diagonal. At this early stage of research, we do not want to exclude the possibility that similar shapes – e.g., a linear arrangement of three points with one point above or below – will also be prototypical. We expect to find an increasing number of prototypes as numerosity increases.

Since the number of prototypes for any given numerosity is greater than one, the partitioning of the similarity space cannot be made using the set of prototypes, but rather a set of subsets of prototypes. More precisely, the solution is to use clusters of points $\{\{p1_1, \ldots, p1_n\},\{p2_1, \ldots, p2_n\}, \ldots,\{pn_1, \ldots, pn_n\}\}$ instead of a set of points $\{p1_1, \ldots, p1_n\}$. As Douven et al. (2013) have shown, the resulting tessellation blurs the boundaries between prototypes – aligning well with the approximate nature of ANS-generated representations (Fig. 10.16).

Of course, this is a simplified picture of ANS representation. We assume that the boundaries between successive numerosities will become thicker as numerosity increases further. We also assume that, for large numerosities, the boundaries can overlap in more than one area: an outcome consistent with the logarithmic distribution of the compressed numerical line of ANS representations (Fig. 10.17).

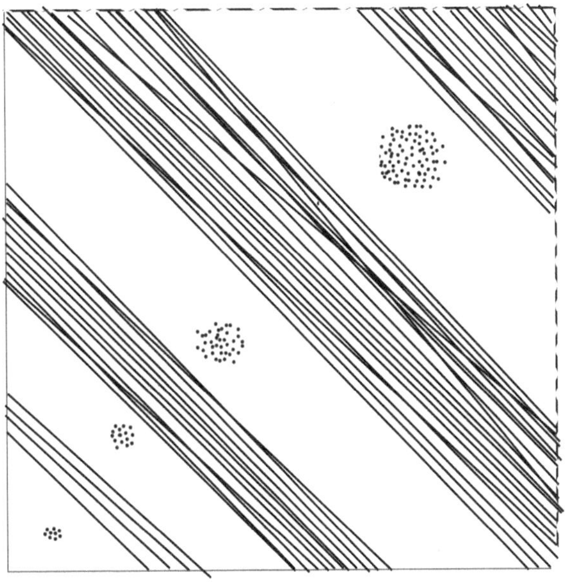

Fig. 10.16 Structure of ANS representations for successive numerosities. Dots represent clusters of generators, lines the blurred boundaries between successive numerosities

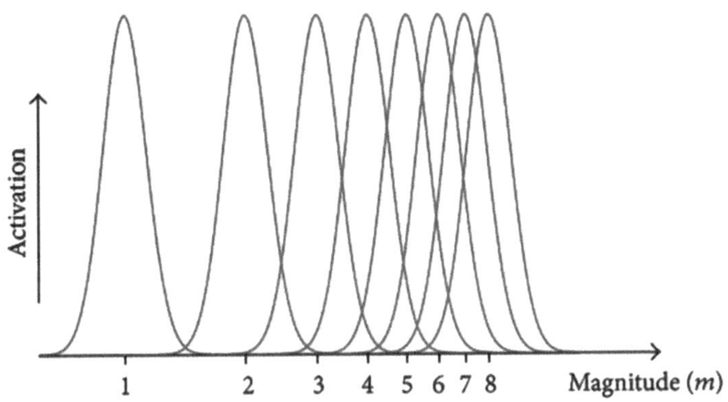

Fig. 10.17 Logarithmic distribution of ANS representations

10.4 Conclusions

The proposed conceptual-spaces model of ANS reflects the specific nature of the system modeled, accounting for size and distance effects in accordance with Weber's law. The model is sensitive to number-related cues. It can, we believe, be easily reconciled with Dehaene and Changeux's sub-symbolic model, while avoiding its numerous problems. Further investigation into the exact nature of the ANS is clearly needed: in particular, taking into account empirical data on inputs, corresponding to prototypical representations.

Acknowledgements We are very grateful to Peter Gärdenfors, as well as to the anonymous reviewers for the valuable comments and helpful remarks. Aleksander Gemel gratefully acknowledges financial support by the *Approximate Number System in Conceptual Spaces* project funded under the Grant of the Dean of the Faculty of Education Science University of Lodz (B1611800000228.01).

References

Banks, M. S. (1980). The development of visual accommodation during early infancy. *Child Development, 51*(3), 646–666.

Burr, D., & Ross, J. (2008). A visual sense of number. *Current Biology, 18*(6), 425–428.

Carey, S. (2001). Cognitive foundations of arithmetic: Evolution and ontogenisis. *Mind & Language, 16*(1), 37–55.

Carey, S. (2009). *The origin of concepts*. Oxford: Oxford University Press.

Carey, S., & Sarnecka, B. W. (2006). The development of human conceptual representations. In M. Johnson & Y. Munakata (Eds.), *Processes of change in brain and cognitive development: Attention and performance* (Vol. XXI, pp. 473–496). Oxford: Oxford University Press.

Dehaene, S. (1997). *The number sense: How the mind creates mathematics*. New York: Oxford University Press.

Dehaene, S., & Brannon, E. M. (2010). Space, time, and number: A Kantian research program. *Trends in Cognitive Science, 14*(12), 517–519.

Dehaene, S., & Changeux, J. P. (1993). Development of elementary numerical abilities: A neuronal model. *Journal of Cognitive Neuroscience, 5*, 390–407.

Dobson, V., & Teller, D. Y. (1978). Visual acuity in human infants: A review and comparison of behavioral and electrophysiological studies. *Vision Research, 18*(11), 1469–1483.

Douven, I., Decock, L., Dietz, R., & Egré, P. (2013). Vagueness: A conceptual spaces approach. *Journal of Philosophical Logic, 42*(1), 137–160.

Feigenson, L., Carey, S., & Hauser, M. (2002a). The representations underlying infants' choice of more: Object files versus analog magnitudes. *Psychological Science, 13*(2), 150–156.

Feigenson, L., Carey, S., & Spelke, E. (2002b). Infants' discrimination of number vs. continuous extent. *Cognitive Psychology, 44*, 33–66.

Feigenson, L., Dehaene, S., & Spelke, E. S. (2004). Core systems of number. *Trends in Cognitive Sciences, 8*(10), 307–314.

Gärdenfors, P. (2000). *Conceptual spaces: On the geometry of thought*. Cambridge, MA: The MIT Press.

Gebuis, T., & Reynvoet, B. (2012a). The interplay between nonsymbolic number and its continuous visual properties. *Journal of Experimental Psychology: General, 141*(4), 642–648.

Gebuis, T., & Reynvoet, B. (2012b). Continuous visual properties explain neural responses to nonsymbolic number. *Psychophysiology, 49*(11), 1481–1491.

Gemel, A., & Quinon, P. (2015). The approximate numbers system and the treatment of vagueness in conceptual spaces. In L. Lukowski, A. Gemel, & B. Żukowski (Eds.), *Cognition, meaning and action* (pp. 87–108). Kraków: Jagiellonian-Lodz University Press.

Goodman, N. (1976). *Languages of art: An approach to a theory of symbols*. Indianapolis: Hackett.

Halberda, J., & Feigenson, L. (2008). Developmental change in the acuity of the "number sense": The approximate number system in 3-, 4-, 5-, and 6-year-olds and adults. *Developmental Psychology, 44*(5), 1457–1465.

Leibovich, T., & Ansari, D. (2016). The symbol-grounding problem in numerical cognition: A review of theory, evidence and outstanding questions. *Canadian Journal of Experimental Psychology, 70*(1), 12–23.

Leibovich, T., & Henik, A. (2013). Magnitude processing in non-symbolic stimuli. *Frontiers in Psychology, 4*, 375.

Leibovich, T., Henik, A., & Salti, M. (2015). Numerosity processing is context driven even in the subitizing range: An fMRI study. *Neuropsychologia, 77*, 137–147.

Leibovich, T., Katzin, N., Harel, M., & Henik, A. (2017). From "sense of number" to "sense of magnitude": The role of continuous magnitudes in numerical cognition. *Behavioral and Brain Sciences, 40*, e164. https://doi.org/10.1017/S0140525X16000960.

Libertus, M. E., Feigenson, L., & Halberda, J. (2011). Preschool acuity of the approximate number system correlates with school math ability. *Developmental Science, 14*(6), 1292–1293.

Lourenco, S. F., & Longo, M. R. (2010). General magnitude representation in human infants. *Psychological Science, 21*(6), 873–881.

Mix, K. S., Huttenlocher, J., & Levine, S. C. (2002a). Multiple cues for quantification in infancy: Is number one of them? *Psychological Bulletin, 128*, 278–294.

Mix, K. S., Huttenlocher, J., & Levine, S. C. (2002b). *Quantitative development in infancy and early childhood*. Oxford: Oxford University Press.

Nieder, A., & Dehaene, S. (2009). Representation of number in the brain. *Annual Review of Neuroscience, 32*, 185–208.

Slusser, E., & Sarnecka, B. W. (2011). Find the picture of eight turtles: A link between children's counting and their knowledge of number-word semantics. *Journal of Experimental Child Psychology, 110*, 38–51.

Spelke, E. (2000). Core knowledge. *American Psychologist, 55*(11), 1233–1243.

Starkey, P., & Cooper, R. G. (1980). Perception of numbers by human infants. *Science, 210*, 1033–1034.

Tversky, A. (1977). Features of similarity. *Psychological Review, 84*(4), 327–352.

Verguts, T., & Fias, W. (2004). Representation of number in animals and humans: A neural model. *Journal of Cognitive Neuroscience, 16*(9), 1493–1504.

Xu, F. (2000). *Numerical competence in infancy: Two systems of representation*. Paper presented at the 12th Biennial International Conference on Infant Studies, Brighton, UK, 16–19 July 2000.

Xu, F. (2003). Numerosity discrimination in infants: Evidence for two systems of representations. *Cognition, 89*, B15–B25.

Xu, F., & Spelke, E. S. (2000). Large number discrimination in 6-month-old infants. *Cognition, 74*, B1–B11.

Mauri Kaipainen • Frank Zenker
Antti Hautamäki • Peter Gärdenfors
Editors

Conceptual Spaces: Elaborations and Applications

Springer

Editors
Mauri Kaipainen
Perspicamus LTD
Helsinki, Finland

Antti Hautamäki
Department of Social Sciences
and Philosophy
University of Jyväskylä
Jyväskylä, Finland

Frank Zenker
International Center for Formal Ontology
Warsaw University of Technology
Warsaw, Poland

Peter Gärdenfors
Cognitive Science
Lund University
Kungshuset, Lundagard
Lund, Sweden

Synthese Library
ISBN 978-3-030-12799-2 ISBN 978-3-030-12800-5 (eBook)
https://doi.org/10.1007/978-3-030-12800-5

This Springer imprint is published by the registered company Springer Nature Switzerland AG.
The registered company address is: Gewerbestrasse 11, 6330 Cham, Switzerland

Kolibri 360 GmbH
Ferdinand-Jühlke-Straße 7
99095 Erfurt, Deutschland
produktsicherheit@kolibri360.de